The Impossible Happens

A Scientist's Personal Discovery of the Extraordinary Nature of Reality

The Impossible Happens

A Scientist's Personal Discovery of the Extraordinary Nature of Reality

Imants Barušs

Winchester, UK
Washington, USA

First published by iff Books, 2013
iff Books is an imprint of John Hunt Publishing Ltd., Laurel House, Station Approach,
Alresford, Hants, SO24 9JH, UK
office1@jhpbooks.net
www.johnhuntpublishing.com
www.iff-books.com

For distributor details and how to order please visit the 'Ordering' section on our website.

ISBN: 978 1 78099 545 8

A CIP catalogue record for this book is available from the British Library.

Design: Stuart Davies

Printed and bound by CPI Group (UK) Ltd, Croydon, CR0 4YY

Contents

Other Books by Imants Barušs

The Personal Nature of Notions of Consciousness: A Theoretical and Empirical Examination of the Role of the Personal in the Understanding of Consciousness

Authentic Knowing: The Convergence of Science and Spiritual Aspiration

Alterations of Consciousness: An Empirical Analysis for Social Scientists

Science as a Spiritual Practice

I show you how deep the rabbit-hole goes.[1]

For Jennifer and all the rest of my students who do not want me to stop talking.

Acknowledgments

I thank Jeanette Wayne, Andrea Markland, Jennifer Oostyen, Monika Mandoki, Rebekah Despard-Young, Ian Brown, Karen Fricke, Cynthia Read, Randal Williams, Kirsten Berner Veiledal, Aucher Serr and Jeffrey Mishlove for reading the manuscript and providing me with feedback. I also thank my research assistants Shannon Foskett, Lindsay Buckingham, Basia Ellis, and Carolyn van Lier for unearthing, retrieving, and organizing various library resources; and Shannon for critiquing, fact-checking and editing the manuscript. I thank Cherie Barušs for her comments about the publisher's contract. None of my readers or assistants are responsible for any errors that might yet remain in the manuscript, nor do any of them necessarily endorse the ideas that I have expressed in this book. I am grateful to King's University College at The University of Western Ontario for research grants that have made the production of this book possible, and Medical Technology (W.B.) Inc. for financial support that, together with research grants and other forms of support from King's, have made possible some of the research described in this book. And, finally, I thank John Hunt for his enthusiasm for this book and for his and his staff's dedication to publishing and promoting it.

Chapter 1

Prologue

There were only some 20 minutes left in the class about consciousness that I was teaching, so I thought that I should just get the students to write their in-class essays and leave. But then I remembered that I had wanted to introduce the notion of morphogenetic fields and asked them if they wanted to hear about it. Jenn said 'yes' right away. 'I just want to keep listening to you talk', she said, by way of explanation. When I was collecting their essays as the students were leaving, I joked with her about her desire to have me keep talking. But Jessica and Fernando, who were in the immediate vicinity of the doorway, did not think it funny, and echoed her sentiment. Of course, there are keen students in every class, and it is they who enrich the classroom experience for all of the students as well as the professors. But Jenn's comment stayed with me.

I have known for a long time that there is something inherently mesmerizing about the subject matter of consciousness, particularly when anomalous phenomena are taken into account. Not just mesmerizing; there is something hopeful about it. In studying consciousness, scientists, long trapped in the bleakness of scientific materialism, have begun to emerge into fresh ways of thinking about reality that have unexpected implications. So I started thinking about writing another book about recent advances in consciousness studies. I went so far as to contact the acquisitions editor of one of my previous publishers who said that she would like to see such a book.

Then I had a dream on the night of January 20, 2010:

Dream: *I met up with one of my mentors in my dreams. She was asking me about my work. I was telling her that I needed to write a* *new book and that an editor from one of my previous publishers wanted to see it. She said that that was a good publisher. I remembered that I already had a book, already written, that needed to be published. And it occurred to me that I could ask the editor if she wanted to publish that book.*

The mentor in my dream was a professor of psychology in real life who has always encouraged my work. The editor in the dream was the editor who had requested to see the proposed book. What surprised me was the suggestion in the dream that there was a book that I had already written that needed a publisher. I use the expression *dream architect* to refer to whatever psychological process it is that constructs the contents of dreams. In this case, rather than just assume that the dream architect was mistaken in saying that I had already written a book, I tried to think of what that book could be. How could that make sense? And then I thought of something.

I like to read books in which the author gives an account of what happened to her, particularly books in which the author finds something unexpected that opened up her ideas about the nature of reality. That inspires me. Yet the books that I had written recently had all been academic books in which I had only occasionally mentioned my own experiences. Perhaps it was time to change that and to write an account of my own adventures. In particular, I thought that it would be important for me to chronicle some of the events that had occurred for me that had led me away from materialist beliefs to the recognition that there are remarkable aspects of reality that are not captured in our ordinary interpretations of it. For instance, over the course of several decades, I had realized that my dreams had revealed things to me that had been helpful for charting my course in life. Often these dreams had been precognitive, in that they had showed me future events before they had occurred. I felt that I should write about these transitions before I lost interest in them as I became acclimated to an expanded view of reality. Indeed, I have already written about such experiences, given the 40 volumes of my diary along with various notebooks and an autobiography in which I have kept track of my self-transformation. In that sense, the book of my own adventures would be a book that had already been written. And this dream, with its interpretation, is an example of the way in which a dream can reveal something practical to a person. In this case, reflecting upon my dream changed the type of book that I ended up writing.

Have you ever had unusual experiences occur to you that changed the way that you thought about reality? One needs to be careful in that it is easy to be deluded about what we think is real. There are lots of people with poor reality testing who think that impossible things have happened to them when they have not. One of my first teaching assignments, while still a doctoral student, was to teach a course in psychopathology. During that time I became quite familiar with the variety of ways in which our versions of reality can become distorted. So, in this book, I am not talking about the phenomena associated with psychiatric conditions. The relevance of this subject matter to mental illness would require a separate discussion. But one also needs to be careful not to deny the occurrence of extraordinary events when there is good reason to think that something unusual has transpired. It is equally pathological to defend against such events when they really do happen. Clearly, good discernment is necessary.

In this book I trace my dawning realization that sometimes seemingly impossible things happen. I start by going back in time to some of my first precognitive dreams and show how I came to realize that they really were precognitive. Then I talk about remote healing, whereby I try to heal people at a distance, and talk about a formal experiment that I conducted to see if I were having any effects. While in the midst of my healing experiments, doctors found a lesion the size of a lemon in the middle of my liver, creating a considerable amount of consternation for me. I describe the dreams that helped me to negotiate my way through that health crisis. Confronting one's own death as I did and, no doubt, many readers have, leads to questions about the possibility of life after death. I recount some of the experiences that I have had that are relevant to any discussion of the survival of consciousness. At the end of the book, I consider a radical change of perspective prompted by the synergistic impact of the various seemingly impossible experiences that make up its substance.

I invite readers to come on this voyage of exploration with me. Perhaps I will say something that resonates with you. Or something that turns out to be useful for shaping a more harmonious life for yourself or others. But I want you to critically evaluate what I have written, rather than just adopting any of it without reflection. And I want to make it clear that none of what I say is to be taken as advice. This is just a narrative that can help to inform you of the possibilities of what reality could be like. And each person needs to use the resources at her disposal to make her own decisions in light of the circumstances in which she finds herself. However, it is my hope that you gain a more hopeful attitude toward any difficulties in your life in the course of reading this book. And that you enjoy listening to the story.

Chapter 2

Precognitive Dreaming

Sometimes people are hit over the head with something seemingly impossible that happens to them and their world is turned upside down in an instant. That is not what happened to me. For me there was a gradual accumulation of evidence that became so overwhelming over time that it would have been pathological to deny it. I am a logical creature, both by nature and by training, having studied advanced logic in the course of my MSc degree in mathematics, and it was logic that forced me to acknowledge that the world was more interesting than I had supposed it to be. Initially, the accumulation of evidence occurred over the course of several decades as a result of analyzing my dreams. I found that I had dreams depicting events that would occur for me in the future. In time, I became so used to the precognitive nature of my dreams that I just assumed that whatever occurred in my dreams was a depiction of possible future events. And then I started to learn to change what happened in the future so that it no longer corresponded to my dreams.

How is it possible to anticipate something that is going to happen in the future? Materialists usually believe that the world is completely determined,[2] so there is no possibility of its deviating from its set course. This idea was famously formalized in Pierre-Simon Laplace's contention that we could know the past and future if we could but know all of the forces operating in the present.[3] Of course, in such a completely determined scheme, the knowing itself would also be determined, as would anyone's insistence that such knowing was or was not possible. So, for a materialist, that there could be experiences anticipating future events is at least possible in principle, although I have found that most materialists consider such an idea to be preposterous.

With the advent of quantum theory, the older billiard-ball model of reality has been superseded by a stochastic interpretation so that the future is ultimately undetermined until such time as it is observed.[4] It is difficult to see how any future event could be reliably predicted in such a scheme whatever one's ontological persuasions. So we do not really have good ways of conceptualizing how we could know what will occur in the future. I think that part of the problem lies in our lack of understanding the nature of time and, as that understanding improves, perhaps so will our ability to comprehend precognition. The point is that our theories of reality need to account for the evidence and, when they cannot, then we need to develop new ones.

As a young adult, when I first noticed that I was having precognitive dreams, I thought that they must be coincidences. As I continued to examine my dreams over the decades, I found that the coincidences were often reliable predictors of future events. And then I found that I could sometimes change those future events to ones that were more to my liking. So let me start with some of the earlier dreams.

Dreaming the future

The earliest precognitive dream that I can recall occurred sometime in early December while I was a student at the University of Toronto; a few months after defecting from the Engineering Science program. The dream occurred before I had started keeping a diary. However, I had written down the dream a few years later.

> ***Dream:*** *I dreamt that it was dark outside and that I was on a bus with a number of other people. I was sitting in the back seat when a man got on and came all the way to the back of the bus and sat down*

> *beside me. I tried to ignore him. But he crowded me, clutching at me with his hands. I pushed him away, but was too polite to insist that he leave me alone. After a while, he would be at it again, and I would half-heartedly push him away again. This pattern repeated itself a number of times before I finally got up and got off the bus. I stood by the side of the road and watched the bus disappear in the dark.*

I thought about the dream when I woke up, could not make anything of it, and so I forgot about it.

About three weeks later I left for a convention. There was a large number of us going to it and so a bus had been chartered so that we could go down as a group. By the time I got to the bus, most of the seats had been taken. I took one of the seats that had been left at the very back of the bus.

But I was not the last one on. Angelo, as I shall call him here, got on after I did. He came to the very back of the bus and sat down beside me.

As the ride got underway, Angelo started asking me penetrating questions about some of the topics that were to be discussed at the upcoming convention. I did not particularly want to answer his questions. I did not feel that it was appropriate for me to do so. But rather than telling him to leave me alone, I made my answers as brief as possible, in the hope that he would give up on me. And I pointed out the superficial nature of his understanding. Angelo was not put off by my responses but took my remarks in stride. After thinking about what I had said he would approach me with another question. Again, I would be as terse as possible. This awkward pattern repeated itself through the wee hours of the night. It was only days later, when I saw him departing the convention in another vehicle, that I recalled the dream and made the connection.

It is interesting to note that there are both literal and representational elements in the dream. The most striking literal element is the dream sequence in which I sit down at the back of the bus and someone gets on and also comes to the back of the bus to sit beside me. I had not expected to be sitting at the back of a bus at the time that I had the dream. The physical clutching in the dream could be symbolic of the persistent questioning in real life. The feeling of having that attention directed toward me was the same in the dream and real life, as was my feeling of politeness. Getting off the bus and watching it depart in the dream was representational in that it signified being free of Angelo's presence. The fact that I saw Angelo leave in a vehicle is a literal link to the dream images. So, there are strong correspondences between the dream and waking life. Just a coincidence? Perhaps.

Dreaming about eating

Shortly after the back of the bus incident, I chose to begin a vegetarian diet. It was not quite a year after that, that I had another precognitive dream. I dreamt that it turned out that I had been eating meat all along. However, in the dream, that realization did not upset me. Upon awakening, I thought that my reaction in the dream had been odd. Had it been true that I had actually still been eating meat, I figured that I would have been furious. It was a few days later that someone told me that cheese contained rennet which is obtained from the stomach linings of ruminants. I had been eating cheese which contained rennet, and hence I had been eating meat. But just as in the dream, I was not upset by this knowledge. I realized that I could not have known that. But afterwards, I always checked the cheese I ate for animal rennet. Eventually, having become convinced of the ostensible benefits of a plant-based diet,[5] I stopped eating dairy products as well as meat and became vegan.

The apparent specificity of the eating meat dream somewhat surprised me. But I did not think much of it at the time. The next series of eating dreams occurred about 35 years later when I was used to being alerted to whatever was going on with me by my dreams. This time I tried to figure out what it was that could be giving rise to them.

A few weeks before becoming vegan, I had stopped eating gluten and noticed some beneficial changes to my health. Gluten consists of proteins found in cereal grains such as wheat, rye, and barley. I thought I had eradicated all sources of gluten, but then had the following dream on the night of August 10, 2007:

> **Dream:** *I was playing soccer with whoever was in the street. That led to playing with some strangers who made some food. It was Indian food, vegetables rolled up in some sort of pastry. I was eating these when I realized that I can't eat gluten, so I stopped halfway through one of the pastries.*

Again, a few nights later on August 14, 2007:

> **Dreams:** *I was in a kitchen, perhaps my sister's although it looked like mine. My sister had an old loaf of egg roll bread in a cupboard below the counter. I took a piece of it and started eating. And, while I had my mouth full with the bread, someone pointed out to me that I shouldn't be eating gluten and I realized that the bread contained gluten. I wasn't about to spit out all that was in my mouth. I knew I could absorb a bit of gluten without much harm. But not to eat any more of it.*

I kept scrutinizing the ingredients lists of everything I was eating trying to figure out where the gluten could be coming from and just could not find anything that I was eating that had gluten in it. I wondered if the gluten were a symbol of something toxic going into my body that should not be there, such as stress hormones that were being released as a result of excessive anxiety. But then I had another dream on the night of August 28, 2007:

> **Dream:** *At some point I realized that I was eating something that*

had a pastry shell. I thought, surely this doesn't have any gluten in it, but then I realized that perhaps it does and that I had better stop eating it.

And again on the night of September 7, 2007:

Dreams: *There was something in my dreams about proof of unusual things, but I'm not sure what that was. However at the end, I bit into a hot dog bun before realizing that it was made of white flour which contained gluten and so I managed to fish out almost all of it from my mouth.*

I just could not figure out what the gluten was supposed to represent. Along the previous anxiety theme, perhaps the gluten in my dreams represented worrisome thoughts that had been poisoning me. In case the dream had a more literal meaning, I continued to scrutinize everything I was eating even though I thought I had checked anything suspicious multiple times.

Finally, in the course of my obsessive checking, I read the label on the bulk bin from which I was obtaining couscous that I was buying. At one point, months after going gluten-free, I had started eating couscous for breakfast just for the sake of variety. I had been under the mistaken impression that couscous was a grain, and since it was not on any of my lists of prohibited grains for those with gluten sensitivity, I thought that it was safe to eat. I was puzzled when I read the label on the bulk bin, but upon checking it out, realized that couscous was actually pasta made from wheat in the form of small granules. I had been eating gluten after all! So I stopped eating couscous. And the gluten dreams ceased.

Dreaming about academic publication

By the time I had the gluten dreams, I had already realized that my dreams could alert me to things that I needed to know. Let me backtrack to the dreams that allowed me to figure that out. Those occurred in the context of academic publication. So let me tell that story.

To succeed as an academic, it is important to publish papers and books. Typically, a professor would write a paper, send it to an editor of an academic journal, the editor would send it out to one or more referees to determine if the paper was worthy of publication, and then inform the author of the decision. The paper could be rejected outright, it could be accepted with revisions that the author was required to make, or, rarely, it could be accepted outright. The process of evaluation is an imperfect one, and, as far as I was concerned, essentially a lottery. What I gradually realized was that I would sometimes dream about whether or not my writing would be accepted for publication before I received the letter informing me of the decision from the editor.

As a doctoral student, attending a university in a small city in the Canadian prairies, there was not much to do for amusement, so together with friends, we played hockey pretty much every day. Road hockey on the streets where we lived, floor hockey in the university gymnasium, and ice hockey on open air ice rinks that the city provided during the winter. Dream researchers have established that much of the contents of our dreams are based on events in our lives, although there do not appear to be any straightforward rules for how those contents show up in dreams.[6] I like to think of these contents drawn from experience as the furniture of a room arranged in a specific configuration by the dream architect in order to convey particular meanings to the dreamer. Because of all the hockey that we played, I would be inclined to dream about hockey, but that did not preclude my dreams from being meaningful.

In fact, it was natural for me to make an association between playing hockey and my academic activity. I noted this in the journal entry of January 24, 1988:

> **Dream (in the morning):** *I was playing hockey (road hockey) and scored a lot of goals. I was playing well. Academically things are moving ahead.*

It was sometime shortly after that that I think that I first made the specific association of scoring goals with the publication of papers.

As a doctoral student, while reading a book in which the authors developed a model of Edmund Husserl's characterization of conscious mental acts using formal classical logic,[7] I realized that such modeling could be done more precisely using Grothendieck topoi, which are particular mathematical constructions that can be used for understanding some types of non-classical logics. I pointed this out to one of my professors who told me to write that up and send it off for publication. I did so, with disastrous results. I sent the paper to a philosophy journal from which it was promptly rejected with the acerbic criticism that not only was my paper unacceptable, but that the editors could not even understand what it was that I was trying to do in the first place.

Ok. It did not look as though this were going anywhere. However, undeterred, I sent the paper out to another journal, *Husserl Studies*. At least the editors should be able to understand that the paper dealt with Husserl's philosophy. While I was waiting for a reply from the editor, I had a dream. At the time, this dream did not even seem worthy of recording in my diary, however, the images of the dream events are still in my mind.

> **Dream:** *A group of us were playing floor hockey with a flat, orange, plastic puck in the north end of the university gymnasium. I was on the team that was trying to score on a net at the east side of the gym. I ended up with the puck on my stick at a considerable distance from the net but took a shot anyway. It not only went past all of the other players, but sailed into the net about three feet from the ground*

without even being touched by the goalie.

The following is an entry from my diary made on April 16, 1988 subsequent to the dream:

My dream that my shot unexpectedly scored while playing hockey was correct. 'Categorical Modelling of Husserl's Intentionality' was accepted by Husserl Studies *last week. Yes, that was a 'long shot'.*

The letter of acceptance from the editor, dated April 4, 1988, read, in part, 'Both the readers, and I myself, like your article very much, and we intend to publish it in *Husserl Studies*'.[8]

Now this was an interesting coincidence. I not only dreamt that I scored, but the goal was a long shot that went cleanly into the net. After the cursory rejection by the previous philosophy journal, I thought that publication of my paper was a long shot. Yet it was accepted 'cleanly', although I was asked to write a section explaining the mathematics that I was using. Was it possible that the dream really was alerting me to actual events? Despite having had previous precognitive dreams, it just seemed difficult to believe that that could be the way reality worked and that these were not just coincidences.

Following up on insights from a dream

The next time I noticed precognitive dreams about publication occurred when I was writing one of my books. This was the first dream that drew my attention on January 23, 1994:

Waking Dream: *I was driving a new, black Cadillac that I had won in a lottery. A dear friend of mine had bought me a ticket and the ticket had won.*

Douglas Baker, a British mystic with whom I had studied in the mid-1970s, had taught us to pay particular attention to our waking dreams, the dreams that occur just before we wake up, because these were the windows between sleep and wakefulness through which messages from a transcendent realm could enter into our awareness. Whether or not that is true, or true by virtue of my thinking that it is true, since hearing that, I have paid particular attention to my waking dreams. At first I thought that this dream had to do with renewal of my personality, given that automobiles can be archetypal symbols for the personality. However, the fact that the dream involved a lottery suggested to me that it was about publication. And publication about what else, but the book that I was writing, wondering if it would get published and, if so, by whom. If the lottery in the dream pertained to publication, then the message was certainly a positive one. I win the lottery. What is more, it is not just that I win, but I win a Cadillac, symbolic of having the book published by a respected publisher. But I do not buy the ticket myself. A friend buys the ticket for me. So, the message seemed to be that a friend of mine would provide the means for having the book published by a respected publisher. Definitely a great message.

Now here is where dreams cease to just reflect reality, even a future reality, and help to shape it. It occurred to me that, if the dream were to be veridical, in some sense, then I could use the information in the dream to try to get the book published. To that end, it was not difficult to establish the identity of the friend in the dream. It seemed to me that the dream character was a professor with whom I was chatting at the time via e-mail about my writing. I figured that he could be the key.

My friend had had a book published, so I thought that perhaps his literary agent could find a publisher for my book. It turned out that the agent was not interested. In the meantime, however, my friend had mentioned my book to the director of an academic book publisher. The director, on the other hand, was interested in my book and asked to see the manuscript once I had finished writing it. I was happy to oblige him. Now this was more than what I had hoped.

But it was not over yet. The book had been sent to two reviewers, one of whom recommended that the book be published and the other of whom did not make a recommendation either way. The director brought the proposal to the board of directors. The day before that meeting, I was told that at that meeting a decision would be made whether or not to publish my book. That night I had a dream:

> ***Dream:*** *I was in a schoolyard playing road hockey. I had the ball. In front of me were two players from the opposite team. I realized that I was far away from the net, so I headed toward it, managing to get past the two players to the front of the net. I crossed in front of the goalmouth from left to right. The goalie moved with me but not quickly enough. There was some space between his left leg and the goalpost. I took a backhanded shot but the ball turned into a wad of paper and missed the post by several inches. No goal. The game was over. It had ended in a tie. But the next thing I knew, I was standing in the schoolyard with a group of other players. Another game was about to begin. There was going to be another chance to win.*

Ok. We were playing hockey, so this appears to be a dream about publication. I get past the two players on the other team, that is to say, the two referees, but I still have the goalie with whom to contend. And, well, this is not good. 'No goal'. No goal means that the book is not published. What did not make any sense was the tie. Either the book would be accepted or it would be rejected. A tie and another game did not make sense. However, I took the meaning of the dream to be that my book would not be published. When I went in to school that morning, I told one of my colleagues about the dream, explaining my dejection at my anticipated rejection. By that time, I trusted the correspondence between dream images and actual events enough to think that my dreams really were showing me what was happening in my life at least some of the time.

The following day I found out what had happened. Apparently the meeting had been going well until one of the members of the board had come in late and wondered whether my manuscript could be considered to be 'new age' writing. A discussion had ensued resulting in a decision to send the manuscript to a third referee to settle the matter. As I wrote in my diary:

> *Well. That is interesting! Tabled. Just as in the dream, the game ended in a tie and immediately another game began.*

There was a clear correspondence between the events in the dream and what occurred in waking life. In particular, a game ending in a tie with another game to be played showed up in waking experience as a stayed decision and consultation with a third referee. The unlikelihood of such a correspondence by chance convinced me to a considerable degree that my dreams were alerting me to actual future events.

A changed outcome is reflected in a precognitive dream

Some of my research, which started during my doctoral studies under the supervision of Robert Moore, has been concerned with beliefs about consciousness and reality of scientists and scholars who could write about consciousness in the academic literature. In particular, we found that beliefs ranged from materialist to transcendent beliefs. For materialists, everything is made up of matter; there is nothing else. For those who are somewhat transcendent, what we called the 'conservatively transcendent', there is an essential ingredient, namely mind, that exists alongside physical reality. For those at the extreme transcendent pole of the material-transcendent spectrum, what we called 'extraordinary transcendence', there is nothing other than a universal consciousness that manifests in different forms, including the physical.[9]

As part of our research, Bob and I conducted a survey of participants at the second Toward a Science of Consciousness conference at the University of Arizona Tucson in 1996. We wrote up the results and sent them to a journal for consideration for publication. The journal editor sent our paper to a referee who wrote back with a long list of complaints about our paper. I was upset when I read the complaints because they appeared to be uninformed and punitive. I made the following remark in my diary on December 8, 1997:

> *I have been upset about the incompetence of the referee for Robert Moore's and my paper 'Beliefs About Consciousness and Reality of Participants at Tucson II'. I have written a stiff reply but had dreams a few nights ago in which I was playing hockey and couldn't score.*

I had not actually written down the dreams in which I could not score while playing hockey, so do not know the timing of those dreams relative to receipt of the complaints by the referee. However, the following dream on the night of December 14, 1997 occurred after I had sent a lengthy rebuttal of the referee's remarks to the editor of the journal, but before I had received a final decision about the publication of our paper:

> ***Dreams:*** *I was playing hockey somewhere in the bushes, just myself against one guy (although there was someone else on his team who wasn't really participating, just watching). I had the puck and was dithering around to the right of the net. Finally I decided that I just needed to make things happen, so I got in front of the net and somehow pushed the puck into the net along with the blade of my stick. I was somewhat surprised at my success, checked to make sure that it really was a goal and then held my stick in there with the puck for a while just to emphasize the goal.*

My analysis at the time:

> *I think that this dream is about the Beliefs paper. Basically, I make it get published. The work I put into writing the response to the referee's comments makes a difference. This is different from the message in my dreams just a few weeks ago where I could not score. In other words, what I do makes a difference. I can create my own opportunities, my own plays.*

And the paper did get published.[10] Apparently my rebuttal made a difference as anticipated by my dreams.

What I started to realize at some point was that there is a two-stage process in the utilization of precognitive dreams. In the first stage, a person convinces herself that precognitive dreams really are precognitive. In other words, some dreams predict that a particular event will occur and then it occurs. In the second stage, dreams predict that a particular event will occur but now, because a person knows that that event will occur, if she does not want it to occur, she can change what happens. I am still just learning how to do this, so I am not always successful at averting an undesirable outcome. Nonetheless, precognitive dreams can allow a person to change what happens in the future, at least to some extent. Ironically, at that point, it looks as though the person is no longer having precognitive dreams! Obviously, this is not the same situation as the person who pays no attention to intimations of the future and thereby has no forewarning of what to change.

Dreams about student projects

Sometimes I am not certain to which paper a hockey dream is referring. This was the case with a waking dream that I had on the morning of October 19, 2000:

> ***Waking Dream:*** *Just before waking up I was playing hockey. I was*

> *doing all right. Then, I really wanted to get another goal. I skated the puck down in the offensive zone but couldn't evade the defender, a woman who tenaciously stayed with me. She managed to get her stick on the puck and tried to clear the puck, but I had reached my stick out around her skates and she fell as she tried to shoot the puck. The puck ended up just sitting on the ice in front of me, but behind her, as she went to the ice. The goalie had also gone down, so I just lifted a backhand shot over the goalie into the net. It was the fifth goal.*

To my mind, the dream clearly seemed to be referring to a paper being published. There was a senior editor of a journal who was female with whom I was communicating, but reference to a 'fifth goal' did not make sense.

Amidst a fair bit of writing around that time, there was only one paper that had been submitted for publication that had not been accepted. This was a paper based on an undergraduate student project. It had already been rejected for publication by one journal and had received a long list of criticisms from the editors of another. I felt that some of the criticisms were justified, but there was nothing that we could do at this point given that they should have been addressed at the time of data collection, which was long passed by the time that we submitted the paper for consideration for publication. Nonetheless, for all its faults, I felt that other researchers would find the paper of value. Indeed, in the years following its eventual publication, I received requests for reprints from around the world.[11] The dream seemed to be off the mark in that it suggested that the paper would be published. In exasperation, I wrote a strong letter to an editor of the journal saying that there was nothing that we could do about some of the criticisms, but that I felt that the paper should be published anyway.

At some point I had a dream, which I appear to not have written down in my diary. This is the dream as I remember it:

Dream: *I was in the somewhat dark, upstairs corridor of a house. Behind me, over my left shoulder, was a young woman who was with me. I had an ordinary hockey stick, but on the blade was one of the large, felt pucks with a hole in the center with which we used to play floor hockey when I was a child. I went down the corridor pushing the puck with my stick. The door at the end of the corridor was closed. However, there was a space, several centimeters high, between the bottom of the door and the floor. As I came up to the door, I jammed the felt puck underneath it and declared a goal.*

Upon awakening, I felt that the person over the left shoulder was my student, that this was a dream about the paper describing her project, and that forcing the puck underneath the door represented compelling the editors to accept the paper. Quite frankly, the way things had been going, I did not believe the dream and was pretty certain that the link between my dream activities and publication would be broken. However, there is a relevant journal entry for March 16, 2001:

My student's paper was accepted for publication without further revisions. This means that all of my goal-scoring hockey dreams have come true so far. I thought that this would be the first one that didn't.

The next student projects were done by three students, all in one year. As the academic year was winding down, on the night of April 8, 2003, I had the following dream:

Dream: *There was a hockey game being played outdoors on a sunny afternoon by university students. I was right in front of the opposition's net and recall being able to score many goals although I don't recall specific instances of scoring goals. I also helped some of the students score goals. For the last goal, for example, Johnny from Monday night hockey was carrying the puck across the front of the*

> *net. I stripped the puck from him so it drifted behind him. I couldn't quite catch up to it, but there was a vertical road sign with hash marks on it that came up to the puck and stuffed it somewhat ponderously into the open net. There was no goalie. This was the second goal in which one of the road signs had moved up to score. I just don't remember the details of the previous one.*

I felt that all three student projects were of publishable quality, although there were only two goals explicitly scored in the dream. One of the students, working in collaboration with an obstetrician, had done a great deal of work tracking women's psychological well-being before and after giving birth to their children. I was certain that that study would get published and perhaps one of the others. However, the study of mothers' psychological well-being was the third on my list of studies to write up for publication and I never did get around to it. The other two were written up and did get published.[12,13]

Ok. So I had figured out that dreams could show me what was going to happen in the future. I also realized that if my dreams were alerting me to something to be avoided, then I could sometimes change it. But if the outcome were a positive one, then I had the belief that my dreams were showing me the optimal course of action that I should take. I had assumed that the dream architect was telling me what I *should* be doing. But maybe there is no 'should.' The series of dreams in the next section is noteworthy because I deliberately chose not to accept the opportunity that my dreams had forecast but to ask for an even more desirable outcome. This is what happened.

Choosing a more desirable outcome

I had written another book and sent it to several publishers when, on the night of September 14, 2004, I had the following dream:

> ***Dream:*** *I was sitting with someone in a cafeteria downstairs somewhere. I had bought a Cash For Life lottery ticket. I scratched the lower field first and then the field just above it. There were other fields but I didn't scratch those. Somehow the ticket was upside down, but I could see that I had won in the upper field. I was hoping that I had won $10,000. The winning line started with 'D:' I could read what it said. It said $3217. I was thinking in terms of the debt on my line of credit. I realized that it would not erase it, but would make a good dent in it. I wondered if the ticket had expired. But when my companion and I turned it over, we saw that the expiration date was October 17. The ticket was good.*

This dream suggested to me that the book would be published but that it would not do as well as I hoped. I thought that the amounts of money could represent advances against royalties or perhaps total copies sold over the publication lifetime of the book. Having an actual date was interesting. I interpreted that to be the date on which the book would be accepted for publication. October 17 was about a month in the future and I did not want to wait that long. But I had little confidence in the date. To think I could dream the date on which the book would be accepted for publication seemed a bit far-fetched even given that I knew by that time that my dreams could predict future events. I also found it interesting that the ticket was upside down. I interpreted this to mean that the dream was showing me what was going to happen but that I could not see that yet in my ordinary waking state.

October 17 came and went and nothing happened. Just some rejections. One of the publishers had apparently lost the manuscript. I sent him another copy and eventually he got back to me to say that he was not interested in publishing the book. The following spring, I decided to rewrite the manuscript. Then, on May 29, 2005, I had a dream during a nap:

Dream: *During my nap I had a dream in which I was outside in the dark, taking shots at a goal with a couple of other people. There was a goalie in goal, but then he was replaced by a friend of mine. At that point there was only one other player who passed me the puck. The puck had changed into a square empty tank of some sort that rolled as it came toward me. I managed to get it with my stick, dribble it back and forth a bit as I approached the goalie, and then shoot it at the goalie, quite hard but not as high as I had hoped. It went decisively over the goalie's bent left leg.*

This is interesting. The goalie is replaced with a friend of mine. When I thought about it, I did have a friend who was a publisher. I asked him if he would be interested in publishing the book. After some discussion about the matter he finally said yes. So that dream did come true. The book was accepted for publication just as the empty tank did go into the net. And, in the dream, the tank barely cleared the goaltender's leg, just as in waking life my friend only agreed to publish the book reluctantly after other alternatives had been exhausted. What I found noteworthy, was that my friend agreed to publish the book on October 17, a year and a month after my lottery dream with October 17 as the expiration date of the lottery ticket.

Because of the coincidence with the dates, there was a sense in which this was the destined publisher for my book. But is that really the way in which reality works? I was grateful that my friend was willing to publish the book, but I was not entirely satisfied. For one thing, he wanted me to do more writing. For another, his company only published on demand rather than having print runs in which a batch of books is published in advance of guaranteed sales. And he was a trade publisher. Book publishing is approximately divided into trade publishing and academic publishing. I was hoping to place that book with an academic publisher if at all possible. I decided to keep looking and to try to find another publisher.

A couple of weeks later, on the night of November 6, 2005, I had the following dream:

> ***Dream:*** *I was playing hockey on a snowy street. It was dark out. I was having difficulty handling the puck because I couldn't seem to focus on it in front of me. I tried to make sure of it and carried it deliberately toward the net. I got in front of the net and took a shot that went high into the net. The goalie was backed right up into the net. Then I took several shots from the side of the net. The puck was a red plastic ball. I shot hard toward the net and the goalie stopped them both times.*

Again, there are some interesting features to this dream. Unlike the previous dream in which the empty can barely clears the goalie's leg, the puck goes straight into the net high up, suggesting stronger acceptance of my work. What was puzzling was the absence of defensemen, or players on the other team trying to stop me. All books are usually reviewed by informed readers before acceptance for publication. Of course, my friend had already accepted it for publication without having it reviewed, and there had been no defensemen in that dream either. And even after the puck has gone into the net, I am still trying to score, though unsuccessfully.

I kept trying to find a publisher for my book, but despite considerable interest, no one else was willing to offer me a contract. The book was too 'scientific' for trade publishers and too 'weird' for academic publishers. Finally, one evening, I sent a one-paragraph e-mail description of the book to another publisher I knew asking if he would be interested in publishing it. In the morning I received an even briefer e-mail from his editor saying that his company would be glad to publish the book. In fact, the book was published without further editing or substantial changes.[14] Even after it had been accepted, yet another publisher went through a review process before eventually rejecting the book.

So, in correspondence with the dream imagery, there was no evaluation. And, as in the dream where the puck sails high into the net without being touched, so in real life my manuscript is accepted enthusiastically without any changes by the publisher. Just as the goalie in the dream kept stopping shots from the side of the net, so in real life a different publisher ended up rejecting the book after it had been accepted for publication.

Over the course of four decades, I have learned several things from these types of dreams. One of them is that my dreams sometimes reasonably reliably appear to show me what is going to happen in the future. In particular they can show me what is going to happen to academic papers and books before I find out about their fate through ordinary means. More than that, my dreams can show me possibilities that I can pursue. With both of the books that I discussed, I successfully found publishers by following the clues in my dreams. Furthermore, once it has become clear that dreams can show us what is going to happen or going to potentially happen in the future, then we can sometimes change the outcome, as illustrated in several of the examples.

So, if we can anticipate the future in our dreams, what does this tell us about the nature of time? How do we need to revise our concept of time in order to accommodate this experiential fact? I usually still think of time as a linear stream from the past to the future because I find it useful to do so. But more and more frequently, I also think of time as a dynamically changing now that embodies both the past and the future. In other words, the past and future are already in the present where they are available to our minds. So, when we dream about future events, we do not need to 'reach across time' since the future is already enfolded in the present.

This way of thinking about time also suggests that the past is not fixed, but that making changes in the present could change events that occurred in the past. I do not know the extent to which such changes are possible, so that remains open to experiment. Some of the healing techniques that I discuss in the next chapter are conceptualized in terms of changing past events as a way of altering the present. The idea is that if we change something, then we change it for all of its temporal duration and not just the future. It is not that a particular condition disappears, rather, it never existed. I do not know if we can actually do that, but I think it is worth thinking about that possibility.

So far, I have talked about precognitive dreaming, mostly in the context of academic publication. Now, academic publication is not a concern for most readers, but matters of health could very well be. So I want to talk about dreams that helped me to negotiate a health crisis. But before doing so, let me say a bit about remote healing, whereby a person tries to heal someone at a distance. Impossible, you say? I carried out a couple of experiments to find out.

Chapter 3

Remote Healing

It was the Wednesday before a final examination for one of my classes on the following Friday when I received an e-mail from a student asking me if she could be exempt from the examination because she had just that day been diagnosed with cancer. I said sure, and would she like me to do some remote healing for her? I had recently read Richard Bartlett's book *Matrix Energetics: The Science and Art of Transformation* in which he provided instructions for techniques whose purpose is to create radical transformation, including healing. Among his examples were twisted spines that straightened out, broken bones that reset, and tumors that disappeared. I liked his idea that we can make up the rules by which reality functions: 'God give me the grace to accept the things that I cannot change. And grant unto me the power to change the things that I cannot accept!'[15] This all sounds impossible, but then, that is what this book is about. I had yet to try Bartlett's techniques, but here was an opportunity to do so. And I was going to do this remotely. That is to say, I would do this while sitting at home in front of my computer with the intention that what I was doing would have therapeutic benefits for my student, wherever she was out in the world.

I was not a novice at remote healing. For a while in the mid-1970s I attended group meditations held by Alma Bell in Toronto. We often did remote healing as part of our meditations. Around the same time, I also learned several remote healing techniques from Douglas Baker.[16] When my father was dying of cancer, I did remote healing for him every morning by directing healing energy toward the heart center which, according to some traditions, is located in an energetic matrix underlying the physical body.[17] One morning as I started my meditation, it seemed as though the energy were not going in, and I found out later that he had already died by then. Although not a novice, I was also not an expert. And I was uncertain that such healing efforts made much difference.

I was also aware of some of the research concerning remote healing, usually in the form of intercessory prayer that is done at a distance with participants not knowing whether or not they are the recipients of prayer. The results of these studies were mixed, with some studies demonstrating effects and others not. One of those studies had been conducted by Randolph Byrd to assess whether intercessory prayer had any therapeutic benefits in patients admitted to a hospital's coronary care unit. This had been a double-blind, randomized experiment in which the patients had been randomly assigned to either a group that received distant intercessory prayer or a control group, with neither patients nor staff knowing to which group a patient had been assigned. Each participant in the prayed-for group had three to seven people praying for her, with those doing the praying having been identified as being 'born again' and having an 'active Christian life'. There were statistically significant differences on 6 of 29 measures of physical well-being, all of them in favor of the patients in the prayed-for group. For instance, 12 patients in the control group required intubation or ventilation whereas none in the experimental group did. The author concluded that there was, indeed, a beneficial effect of intercessory prayer.[18]

There is an interesting study of intercessory prayer that I sometimes get my students to read about. It is interesting because the praying was done 4 to 10 years after the events had occurred that were the target of the prayer. The unwitting participants in the study were 3,393 patients at a university hospital from 1990 to 1996 who had had bloodstream infections. In 2000 they were randomly assigned to either an intervention group or a control group. Those in the intervention group received a 'short prayer for the well being and full recovery of the group as a whole'. Three outcome measures were used of which two were statistically significantly different between the intervention group and the control group. Those in the intervention group had a shorter hospital stay and their fevers did not last as long as those in the control group. The third measure, mortality, was also numerically lower for the prayed-for group, but not statistically significantly so.[19] The results of this study suggest that retrocausal effects could be possible, although there is an alternative explanation that we will consider in a while. At any rate, I was aware of at least a couple of studies showing that remote healing could occur.

Matrix Energetics

What I liked about Bartlett's system, Matrix Energetics (ME), was that he had packaged various healing strategies in a way that made them practical to use without presupposing any special abilities on the part of the person using them. One of the key ideas, which he had taken from Rupert Sheldrake, is that everything we encounter has a morphic field associated with it that structures its manifestation.[20] For instance, cancer has a morphic field that includes various symptoms, medical procedures, likely consequences, and so on. Bartlett's contention was that if we confine ourselves to the morphic field of cancer, then we limit the range of results that are possible. If we want radical transformation, then we need to get out from underneath that morphic field so that other outcomes can occur. One way to do that is to disregard the fact that someone has had the label 'cancer' attached to her and to proceed from there. Another key idea, this time taken from quantum theory, is that reality is in a state of superposition with different alternative versions of physical reality available for manifestation, and that we can choose from among those versions. The idea is to choose states of exceptional well-being. We usually also have the notion that if anything is going to change, such as an unwanted condition, then it is going to require considerable effort on our part. Not so, says Bartlett. We can just imagine that a condition is gone. There are also more elaborate techniques, but they all essentially entail creating changes in our imagination and then releasing the intention for change by 'doing nothing.'[21] Eventually we look to see what has changed, as this confirms a new reality rather than reaffirming the old one.

Let me just elaborate on *doing nothing*. I understand this to be a non-dual state in which, as much as possible, distinctions are temporarily held in abeyance. So, in my experience, at the least, I would be unconcerned about whether or not anything that I was doing was having any effect. Or, if there were to be effects, I do not concern myself with whether or not they would correspond to what I think should happen. In other words, I refrain from trying to push around reality with my mind. Deeper versions of this state, for me, have included a more profound detachment from physical manifestation and, sometimes, feelings of deep peace in a state in which I lose track of where I am, who I am, and what I am doing. Such a state can border on non-dual forms of enlightenment such as those which occurred for the American mystic Franklin Wolff.[22-24] The idea here is to get out of the way so that a deeper process can be initiated that has the ability to transform reality.

Ok. Back to the story. That Wednesday evening, I started in on my student. I held her in a sacred space in which there was no cancer and saw her body effortlessly maintaining perfect health. At one point I did a two-point technique. The idea with this technique is to find a point on the body where one's fingers seem to 'stick' as they pass over that point. One hand is placed there. The other hand goes on a second point on or off the body that seems to hold 'tension' with the first point. Then one simply imagines that the two points are connected and releases any attachment to the outcome. Subsequently Bartlett has simplified the procedure in that one can simply select the points by noticing whichever points seem to be appropriate. I used the first method and found two points on my ergonomic computer keyboard, which stood in for my student's body. I did all of this with a sense of commitment to the process and a considerable degree of absorption in what I was doing. I saw a blue light in the middle of a spiral and wondered if it were a healing light.

The next Friday, I wrote the following in my diary:

> ***Dreams (including waking dreams):*** *Several times I had the impression that I was doing the right things with my student; that what I was doing was working.*

The following Sunday, I did ME for her again. I went to a deep place in which I felt a sense of peace. I saw a constellation of about five blue lights in the upper right hand side of my visual field. Then they were gone, but feelings of peace persisted.

I did not find out what had happened to my student until several weeks later, and some parts of the story only came out after many months when she had adequate time to talk to me. The first evening that I had done ME for her, she said that she had been staggering around for 3 or 4 hours, unable to grasp things with her hands, and could not figure out what was going on with her. Of course, that could likely just have been an anxiety reaction to the news that she had just had. She also told me that when the surgeons had cut her open, they had found that the tumor was smaller than they had expected from their ultrasound and magnetic resonance images. She said that she had also been told that she would not need chemotherapy or radiation therapy, although the physicians ended up giving her radiation treatment anyway. I did not know that anything that I had done had had any effect, but this outcome did make me reflect on the possible efficacy of remote healing.

A week after the second remote healing session for my

student, and on the second day of a four-day water fast, I was doing ME for myself. I held a two-point for 3 or 4 minutes and did not think that anything had happened. Then I remembered something that the Italian psychiatrist Roberto Assagioli had said at one point, namely, that it is precisely when we think that we have exhausted all possible associations with a seed thought in meditation, that we need to apply the will in order to persist.[25] I thought that that advice applied to my situation, so I persisted, and wrote the following in my diary:

> *So I held the points. And, after a while, I saw a chink of light break forth from a hole in some sort of floor that was above me, and it fell on the face of a beautiful woman. She was lying on her back with the top of her head toward me. The light fell in an ovoid shape on the right side of her face. I noticed its beauty. Then I imagined facing her. All was light. She was levitated facing me about 12 to 15 feet away. I thought of the phrase 'healing angel'. It's as though the healing angel had been awakened. Until now, I had wondered why my healing efforts had not gone anywhere, but that potential for healing has been dormant. Perhaps now it is awake.*

I attend some training sessions

I had seen enough to convince me that I should take the next step and participate in an ME training seminar. So it is that I found myself at the Level 1 seminar in Miami which ran over the course of a weekend from Friday evening to late Sunday afternoon. I had convinced one of my former students, who was now practicing as a doctor of natural medicine, to accompany me and to help me evaluate what was going on. Now, there have been organized group activities in which I have participated that I regard as highlights of my life, such as a 2 x 2 Latvian youth camp in New England; a week-long Festival of Esoteric Science and Yoga organized by Douglas Baker in Toronto; an eight-day Psychosynthesis intensive workshop in Berkeley; and a meeting of the Society for Scientific Exploration in Santa Fe, New Mexico. This weekend of ME training is now on that list. It was a fantastic experience.

Whatever I did seemed to work as I practiced on the other participants at the seminar. Most notable, perhaps, was the final person with whom I had a chance to interact. I spent perhaps five to ten minutes playing with her using a technique called 'archetypes' whereby we allow images concerning the person to arise in our minds and work with them in ways that seem appropriate to us. The images I saw pertained to her head and, from what I recall, I was prompted to clear out her forehead with some sort of imaginary corkscrew device. I was cut short by the end of the training period, but my practice partner, whom I had not met previously, nor with whom have I had any interaction since then, said that she could see in focus after I had finished with her; something that she said that she had been unable to do prior to that. I did not know that she had had problems with her eyesight and was somewhat surprised but dubious that anything I had done could have made a substantial difference. I did note the interesting coincidence, though, in that, according to Bartlett, ME was born when a hallucination of George Reeves had showed up in his clinic dressed as Superman and directed his x-ray vision toward a patient's brain allowing Bartlett to visualize healing the girl's eyes. She too had been able to see in focus after his treatment.

I had to leave shortly afterwards to fly back home, so I missed the level two training the following day. However, I did have a hypnagogic image during a nap I took at home on that day. *Hypnagogic images* are images that occur as one is falling asleep and can be quite vivid, as this one was.

> ***Hypnagogic image:*** *I was at Bartlett's seminar but it was in a 'room' above the room where the physical seminar was going on. There were about 3 or 4 angels teaching it and even fewer students than in Bartlett's physical seminar.*

The idea in the image was that there was a space in some other dimension above the physical space in which there was a seminar being held. And that, somehow, my desire to continue the training led to my participation in that alternate venue. I have no idea whether that has any relation to reality but, when I checked with Bartlett's entourage, I received the answer that there was such a version of the seminar.

Experiment I

I decided that it was time to conduct a formal experiment of remote healing based on techniques derived from ME to see if I could determine whether anything appeared to be happening and, if so, whether I could get any clues as to what the mechanisms might be. The protocol for the experiment was straightforward. I would e-mail a participant telling her that I was going to begin a session, indicating the time that I would begin, and ask her to tell me if anything out of the ordinary occurred around that time. Then I would do a session and write up a paragraph describing what I had done. Once I had heard a participant's version of what had happened, I would share what I had written. There were 15 volunteers in the study for whom I conducted a total of 34 sessions over the course of nine months with one to six sessions per person. The length of sessions ranged from 3 to 35 minutes with the average length of session being 18 minutes.

The results were interesting. It appeared as though sometimes something happened when I did a remote healing session for someone. And whatever that was, it seemed to be clearly beneficial, even if, at the time of the session, the person experienced some fatigue. It also seemed that more 'sensitive' people appeared to be more likely to feel something happening during the sessions. I want to comment on the results under three categories: Remote viewing, remote influencing, and interactions at a mythological level. 'Remote viewing' denotes the perception of events at a distance without using the usual sensory modalities. 'Remote influencing' refers to influencing events at a distance without any ordinary physical connection to those events. And by 'interactions at a mythological level' I am talking about interactions with participants that involved mythical images.

Although it was not my intention to accurately *see* anything that was actually happening with a person during the remote healing session, I sometimes nonetheless seemed to accurately pick up on what was going on with her. For instance, Participant 09 said 'I am overwhelmed and awestruck by your reflections, you were extremely accurate with everything', and Participant 14 said 'I was literally overwhelmed. I was overwhelmed by your ability to see.' Specific instances of remote viewing will become apparent in some of the examples later in this chapter.

The following is an example of apparent remote influencing. I had done a session for Participant 03 and, 14 days later, did a second session and asked her for her impressions. This was the response that I received:

> ***Participant 03:*** *I wondered whether there was a sending. I began to suddenly feel lighter, more focused, less distressed. The past few days have been overwhelming. I have felt as if I've entered a desert with no boundaries. I am porous like a sea bed, everything seems to go to the bottom and lots of the stuff just stays there, unsorted, unscanned. It tires me and makes me want to scream stop! I can say that first my mind/emotions, then my body relaxed and just hummed nicely. My focus came back and the dead desert feeling faded. I also decided not to do any work and just happily read. It was a subtle but meaningful change. The feeling was like the last time. A feeling of being supported, resurrected, effortlessly lifted from a state of low energy to a higher more harmonious state.*

Seventy days later, I did a third session for Participant 03 and received this response:

> ***Participant 03:*** *Oh, that's what happened. At around 10 or so, I had just woken up from a nap and wondering whether I could be doing some work when I felt a good energy moving through my consciousness. It felt strong and clear, very different from the murky tiredness that made me want to do nothing after I got home around 7:30. My energy now feels strong. I no longer feel the mind fog and body tiredness.*

Now, this does not prove anything. But it does suggest that remote healing could have an energizing effect, helping the recipient be more focused. In fact, I found that Participants 01, 03, 04, 06, 11, and 12 reported increased energy or concentration, whereas Participants 05, 06, 08, 10, and 15 reported increased fatigue, heaviness, or sleep. Note that the same participant could report increased energy for one session and increased fatigue for another. Here is an example from Participant 10 of enhanced sleep: 'Nothing out of the ordinary happened, other than a very profound sleep. I know it was unusually deep because the cat woke me at midnight.'

On one occasion I was doing a session for Participant 03 when Participant 05 came to mind. I thought 'Why not?' and so for about three minutes I was simultaneously imaging healing both participants. Afterwards, Participant 05 had this to say:

> ***Participant 05:*** *It's funny, I know exactly what I was doing at that time (I was brushing my teeth). Again, I did not really experience anything unusual, however, I did picture you performing ME at that time. I am not sure if this is coincidence or not. Oh, and my cat was uncharacteristically friendly after that time; as a night owl, he normally does not like to cuddle at night!*

Who did what to whom? Was I just 'influencing' Participant 05 or did she 'want some' and drew my attention to her while I was interacting with someone else? She has told me that the effect of the remote healing on her is like that of recharging her batteries: 'Every time you perform a session, I feel like my batteries are recharged again. It is truly incredible!' So did some part of her notice that I was doing remote healing and link in somehow? The technique I was doing at the time is something that I call 'alien head.' I go into a non-dual state of consciousness and 'track changes' with 'automatic' movements of my head. As I explained previously, the purpose of such a state is to access deeper levels of reality from which radical transformation can be initiated. It would be as though a shower of goodies were coming down for a while. Did Participant 05 notice the goodies and butt in so that she could get some? I do not know the answer to that.

Participants' cats seem to be affected by the remote healing. After another session, Participant 05 said that 'again, my cat was uncharacteristically friendly.' Although this does not constitute actual evidence of influence, it does remind me of Keith Harary's apparent effect on a kitten named Spirit during out-of-body experiments at Duke University in 1973.[26] Is it possible that animals, such as cats, are more open to whatever occurs during remote healing?

Mythological interactions

Sometimes the stories of what I did appeared to accurately reflect what was occurring physically or emotionally for a participant, but often those stories were representational and spoke to a mythological level of reality. For example, this is what I said to Participant 03:

> ***Imants:*** *You looked zapped; shivering, cold. When I asked to change that into a good process, the shivering slowed down until you froze completely and then broke apart. Inside was a small, fresh person, vulnerable, but, paradoxically, more robust; better able to cope with the world. Then I threw in everything else that I hadn't already offered that would be beneficial for you. You looked happy. You were*

cleaning the house.

And the participant's response:

> ***Participant 03:*** *'Cleaning the House' sounds like a perfect metaphor for what I am doing in more multidimensional ways than ever. I also have a recurring dream where I move into a new house and discover it is not right; and often I have to 'clean house' before I can stay there. The other day, before your healing, I felt frozen, cold, shivering. I mentioned that I wanted to scream just to break the mode/mood. Then it shifted and I was content. I love the way you describe what you do. Its imagery speaks to me.*

But this process can also work in reverse, where a participant perceives what I am doing symbolically. Here is some of the back and forth with Participant 09 in which there are a number of interesting interchanges.

> ***Imants:*** *While in the non-dual state, I noticed that angels had prepared a module for you. This is an informational/energy package designed specifically for you. I placed it in your lower right abdomen, but it also projected a beam of light into your right foot. It helped to stabilize you. When I looked at you, you looked stronger. However, I felt that you could use some cheerfulness, so I turned on Frequency 16, playfulness. When I looked at you again, there was a superposition of a slow smile, but still some darkness in your stomach. I used the archetypes technique to run a non-dual state into the future and collapse all of that benefit into now. Then I saw you making and eating muffins. That signified normality to me. The non-dual state didn't seem to want to end and seemed to be beneficial for you, so I created an alternate version of myself that would stay in the non-dual state keeping the healing current open until harmony was reached.*

I did know that the participant had baked muffins in the past. And the alternate version of myself is just an imagined self that I set in place and then left behind to function as needed. This is what I heard back from the participant:

> **Participant 09:** *Wow. I am overwhelmed and awestruck by your reflections. I don't know if I should tell you the medical specifics of what I've been through recently. But I'll just say that you were extremely accurate with everything. This morning I had a dream with you in it. During the dream I was at my house, and I came outside to our front porch, and you were sitting in a wooden Muskoka chair, all wrapped up in blankets, snacking away on your cat's vegetarian cat food (? this makes me laugh) and your cat was on your lap. Later in the dream you told me that this cat's name was Louise, and you had her in this little cat-holder, which was kind of like a purse for cats with a strap, and the cat was quite content to be in her holder (upon waking this also made me laugh). I started petting the cat and playing with her. When I came outside I was surprised but happy to see you, and you said that you had just performed a ME session on me. You said that I am very much in need of this healing right now. I was immediately struck by this dream. It was extremely vivid and now reading your description of the session I am certain that it was the alternate version of yourself that I encountered.*

I found this dream to be interesting. I realized that, for me, the cat could represent the magical power of remote healing. And I had actually imagined Lake Louise during the session as a gateway to the non-dual state of consciousness. I pointed this out to the participant.

> **Participant 09:** *I completely agree about your thoughts on cats. My cats have always been very special, safe, positive and healing beings for me, so it doesn't surprise me that healing somehow*

manifested as a cat in the dream. Also, I was going to mention this in the last email but I didn't think it was really pertinent, but now that I think about it, it might be: the cat holder that that cat was in was a rich blue/green turquoise color. The rest of the dream was otherwise quite bland in color. In regards to what has happened to me, I recently lost my first pregnancy. At the same time I found out that I had a cyst on my right ovary. Hence the lower right abdomen and darkness in my belly. In terms of the right foot there is nothing physically wrong with my right foot, but I do very much identify with the feeling of my feet being swept out from under me as of late, from this whole traumatic experience. Further, over the last couple of weeks I have been a veritable muffin factory, almost unconsciously driven to bake, and these kinds of domestic acts like baking have really helped to ground me after what's happened. And I am definitely in need of a good dose of cheerfulness :). I am wondering if the silly aspects of my dream that made me laugh were merely to add a bit of playfulness to my day.

Imants*: I wonder if the whimsical part of you draws out the whimsical part of me. From the moment I first saw the rivers and lakes in Banff National Park I was struck by their turquoise color as much as by their crystal like clarity, so the color is definitely important to me, and seems to cross validate the interpretation of the cat as the healing power associated with Lake Louise.*

Participant 09 *(55 days after previous e-mail to me): My cyst is gone!*

This interchange raises more questions than it answers. What is the significance of interactions at a mythological level? How is it that we can apperceive each other's mental contents? And, perhaps more to the point, does any of this imaginary play have anything to do with actual healing? For instance, did anything that transpired during the session facilitate the disappearance of the cyst or did it disappear by itself? How does any of this lead to healing? But I had seen enough to know that something seemed to be happening, so it was time to introduce a control condition.

Experiment 2

I needed a way to verify that participants were actually being affected by the remote healing rather than just the expectation that something beneficial could happen. Usually this is done by introducing a control condition in which the active ingredient, in this case, actual remote healing, is missing, and then making comparisons between the presence and absence of the active ingredient. I had held off introducing a control condition in the first experiment because I was aware of the work of Bill Bengston. When Bengston treated mice that had been injected with breast cancer using laying-on-of-hands, he found that untreated animals who were part of his experiment were often healed along with the treated animals.

In Bengston's experiments, if left untreated, all of his mice would have died within 14 to 27 days. The laying-on-of-hands was largely successful in treating the mice. For instance, in a series of studies in which either Bengston or skeptical volunteers who had been trained in Bengston's technique did the laying-on-of-hands, there was an 88 per cent success rate with 33 experimental mice. Those are interesting results. But what was perhaps more interesting was that often the cancer in the control mice would remit as well, even though they were never treated. In the same series of studies, 70 per cent of 26 control mice that were at the experimental site also went into remission. However, eight control mice sent to another city all died.[27] I recall once when Bengston was talking about his experiments, he said that without his knowledge students had placed a mouse cage with a mouse in the same lab as the one in which he was doing laying-on-of-hands and that that mouse remitted even though other mice in reasonably close proximity to him who were never part of the experiment were unaffected.[28]

This is a significant observation. It suggests to me that the world is carved up according to meaning rather than physical proximity. In other words, mice that were in some way part of the experiment were affected, even if the investigator himself did not know at the time that they were part of the experiment, whereas mice that were never part of the experiment, were not affected. Of course this raises the question of where the boundaries of the experiment are, particularly since not all control mice remitted. In my case, I thought that if there were to be a healing effect, then it might extend to the participants in a control condition as well as those in an experimental condition. However, I hoped that an actual remote healing session could augment whatever effects were present beyond those that would occur just from being a participant in the study. So I decided to introduce a control condition.

The simplest modification to the previous protocol was to tell participants that I would begin a session at a particular time, but then to toss a coin. If it landed heads, I would proceed as I had previously, by doing remote healing. That would be an experimental session. And if the coin landed tails, I would not do anything further. That would be a control session. In either case I would wait to hear back from participants before telling them whether or not they had been involved in an experimental or control session and, in the event that it were the former, once I had heard from them, I would send them the paragraph describing what had occurred during the session.

In order to have something to measure, I asked participants to respond to three statements on a scale of one to six from 'Strongly Disagree' to 'Strongly Agree' before they knew whether or not they were involved in an experimental or control session:

1. I experienced something unusual at the time of the session.

2. I felt more fatigued than I expected to have been at the time of the session.
3. I felt more energized than I expected to have been at the time of the session.

I chose to measure self-reported changes in energy because these seemed to be the most frequent changes that participants had reported in Experiment 1. It was time to determine whether participants' energy levels really were being affected.

I also filled out two measures for myself, on a scale of 1 to 10, immediately upon completing a session in which I did remote healing:

1. Degree of absorption (steadiness of mind, lack of distraction) from 1 (none) to 10 (complete).
2. Depth of non-dual state (non-duality of non-dual state, degree of release) from 1 (shallow) to 10 (deep).

In other words, how well did I concentrate on what I was supposed to be doing? And how deeply did I seem to enter a non-dual state of consciousness? The idea was to check whether either of these two were factors in any remote influence that I might be having.

There were 22 participants in Experiment 2 with a total of 138 sessions, 72 of which were experimental sessions and 66 control sessions, from May 26, 2010 to May 11, 2012. There were 1 to 11 sessions per participant with a median value of 7. The average length of experimental sessions was 22 minutes.

In all cases the averages for questionnaire items one to three were higher in the experimental condition than the control condition, although not statistically significantly so. However, clearly a participant would not be more fatigued and more energized during the same session, so I looked at the absolute values of the differences between being more energized and more fatigued for the experimental condition (which had a value of 2.08) and the control condition (which had a value of 1.56). The difference between experimental and control conditions only had a likelihood of occurrence by chance of 4 per cent. Similarly, I made a little scale by adding the response of the first questionnaire item to the absolute value of the difference between being more energized and more fatigued and found a difference between experimental and control conditions whose likelihood by chance was about 5 per cent. Those are statistically rare events and suggest that I was having an effect on changes in energy levels of participants.

This was not unexpected, given what I had been reading in the responses from my participants. However, it did mark a milestone for me, in that it was no longer just the case that it seemed as though I were having an effect on participants, but I now knew that I was having an effect as determined by standard statistical analyses of numerical data. Something really was happening. Participants were noting changes to their energy levels as a result of remote healing.

How are we to understand the statistically significant results? Am I really influencing participants in such a way that their energy levels change? I think so, as judged by the previous examples as well as the examples that I will give below. But there is an alternative possible explanation called Decision Augmentation Theory. The idea is that experimenters' decisions about what to do are augmented by remote viewing so that, for instance, experimenters place those with a better prognosis into the treatment group whereas those with a worse prognosis end up in the control group. This would also presuppose a paranormal ability to affect random events (in this case, the coin toss) in the course of randomly assigning participants to conditions. In other words, the idea is that it is the researchers' activities with regard to the logistics of a study and not any actual healing that is considered to account for favorable outcomes.[29]

This is a possible explanation for the effects in the apparently retrocausal intercessory prayer study mentioned previously. In my case, it would mean that I paranormally affected the coin tosses in such a way as to select participants for remote healing who were going to have energy changes during the time of the sessions anyway. However, it seems to me that if I could paranormally affect the coin tosses, then I could also affect the participants. And, as is evident in some of the examples below, some of the effects that I appear to have had are sufficiently unique as to be unlikely to have occurred spontaneously. Or at least, the correspondence between what I was doing and what the participants reported is so striking, that it is difficult to deny the possibility of a connection.

Examples of remote healing

Statistical analyses of data are important for establishing the likelihood that something is actually happening. But I want to give several more examples of correspondences between what I was doing and what participants were reporting. In each case, I will indicate part of what I had written immediately after the session, and then what the participant wrote before the participant had seen what I had written or even knew whether it had been an experimental or control session. In the first example, according to the participant, she did not see my e-mail until the morning after the events she describes had occurred. I will explain what I say in my description, including the ME terminology, immediately following it.

> ***Imants (regarding Participant 11):*** *You were pointing at your wrist. You wanted to get rid of whatever the problem was. What I saw was that the problem is the result of the suppression of energy in your body. There is insufficient energy flowing to the periphery of your body. I asked the question of what it would take to restore the energy and touched the wavefront of the answer. There ended up*

> *being three components: 1. Pet some kittens (they don't have to be yours); 2. Spin with your arms outstretched so that subtle energy is forced down your arms; and 3. Go into the mountains and find a cold stream and stick both hands into the stream. I had and hadn't placed your 'problem' on a clipboard. So I looked at the clipboard again. The problem with your wrist was gone.*

In this case, the participant had told me that she had a cyst on her right wrist and that she was hoping that I could help her to get rid of that, so there is nothing surprising about imagining her pointing to her wrist. With regard to the 'suppression of energy' in the body, I have no idea whether or not that even refers to any actual physical processes. I mean, there are lots of theories about energies in the body but the evidence regarding their specific activity is somewhat patchy.[30] For the purposes of the remote healing, that does not matter. I just use whatever imagery arises for me. I did not know how to fix the problem with energy, but I trusted there was some intelligence somewhere with the answer. That is what I mean when I say that I 'touched the wavefront of the answer.' At that point, I do not even need to know what the answer is. It could be self-implementing. But, in this case, I 'looked', and saw the 'three components'. And the three components are hilarious! I have pondered how such activities could have any therapeutic benefit and it seems to me that, at a minimum, they would throw a person out of her usual routines, allowing something new to show up that could be helpful. Finally, let me explain the statements about the 'clipboard.' If we know what is wrong with someone, we can 'put it on the clipboard', thereby disregarding it. In this case, I both had and had not placed her problem with the cyst on the 'clipboard'. This is a quantum-like activity in that I was imagining both alternatives proceeding simultaneously.

This is what I heard back from Participant 11 after telling her that I would be doing a session for her, without telling her anything further:

> ***Participant 11:*** *I was going thru the week's miscellaneous mail and catching up on reading material. I became aware of a pulsating feeling in the fingers of my left hand (I'm right handed. Yes, it momentarily was a what in the world is this all about concern. But the feeling, shortly, ceased.) What was even more noticeable was my energy level. It seemed to be heightened. I was more alert and focused. I sped thru the week's mail and literature over the next 2 hours. I don't ever recall having this level of concentration available on an end of the week Friday evening.*

The increased energy level with the accompanying ability to focus is not surprising, considering that I had found that fairly frequently and now had statistically significant evidence that such energy changes were occurring during remote healing. What is surprising is the correspondence between what I was doing in my imagination by seeking to find ways to increase the flow of 'energy' to the periphery of the participant's body and the actual sensations that she was feeling. At the time that the participant wrote the e-mail to me, she did not know whether the session had been an experimental or control session, let alone what I had done. It is also interesting that the possibly anomalous sensations occurred in the left hand, whereas it was the right wrist that had the cyst. I do not know at what point the cyst disappeared and do not know if anything I did had anything to do with it, but 20 days after the previous e-mail, I received the following:

> ***Participant 11:*** *I felt we had great success with the ganglion cyst on my right wrist. It is still GONE!*

Here is another example. This is what I wrote after doing a session for Participant 05:

> **Imants:** *When I looked at you, you looked overwhelmed. I felt that something had come up with your health. I used a locator and was led to the lower back of your head. Possibly back teeth or jaw. Possibly temporomandibular joint. I used the Master Corrections Scan. The negative meter was at minus 33 but could be moved up to minus 17 at which point the problem would no longer be a problem. Self, psychic symbols, and body organs came up. I started with body organs since they seemed to be an alternative to the other two. None of the organ systems showed up. I suspected temporomandibular joint or something of that sort. I went with psychic symbols and saw something like floss on both sides of the back of your mouth. I pulled it out and set fire to it; it burned with an iridescent hue and was transmuted into something worthwhile. I went into a non dual state.*

I felt that a health issue had come up for this participant. A 'locator' is just a moving set of three-dimensional Cartesian coordinates that one moves intuitively through a person's body to find the physical source for a problem. The 'Master Corrections Scan' is a level 4 protocol in ME for determining what to do in order to correct whatever shows up in a state of disequilibrium. A 'negative meter' is an imaginary meter from minus 100 at the bottom to 0 at the top and can be used to intuitively gauge the extent of a problem. In this case, the problem was at minus 33, so it was not particularly serious. However, I noticed that it could not be moved to zero, but only up to minus 17. However, at that point, for practical purposes, the problem would have been rectified. 'Self, psychic symbols, and body organs' are specific ME level 4 protocols. The expression 'psychic symbols' refers to a representation of something in a person's body that should not be there. In this case, it had the appearance of dental floss. Once a psychic symbol is located, we are to imagine some means of getting rid of it. I thought I would burn it in my imagination and transmute it into 'something worthwhile'. Then I went into a non-dual state with the intention that that would enhance the effectiveness of whatever else I had done.

> ***Participant 05:*** *As for the session, it feels like it was an actual one. Today, I feel really happy and full of energy despite having written a difficult examination. I have had a lot of neck pain for the past several weeks, and today it seems to be almost gone.*

This response was received from the participant before she knew whether or not I had done an actual session. There are several things that I find interesting here. The first is the confirmation that she had had pain in her neck. The second is that the pain subsided, but not completely, so that there is a correspondence with the change in readings on the 'negative health-o-meter'. The third is the increase in energy and feelings of happiness, which are a common theme in participants' responses. I e-mailed my write-up to the participant and received this in turn:

> ***Participant 05:*** *It is amazing how precise you were with the neck pain. I cannot believe the relief I feel. Whenever you perform these sessions, it completely transforms how I feel. Clarity and happiness would be two words that would best describe this.*

Further discussion with the participant revealed the extent to which she had felt elated for some three days following the session. I had also noted such improvements in mood in speaking to participants at Bartlett's seminars. But if these are actual effects, then the question arises of the mechanism through which they work.

Speculations about the mechanisms of remote healing

When I analyzed the data from my own self-report measures, I found something interesting. There was a positive correlation between my depth of non-duality and participants' feelings of fatigue and a negative correlation between my depth of non-duality and participants' feelings of being energized. This led me to calculate the difference between being energized and fatigued, which ended up having a negative correlation of $r= -.29$ with depth of non-duality. The probability of occurrence of such a correlation by chance is less than 5 per cent. In other words, the more that I had gone into a non-dual state of consciousness the more fatigued participants had felt, and the less I had been able to enter a non-dual state, the more energized participants had felt. How did that make sense?

What this suggested to me was the presence of two separate mechanisms of remote influencing. The more deeply I am able to enter a non-dual state, the more profoundly participants are affected by healing processes that lie outside of myself. The idea is that people become fatigued and 'zone out' while such deep processes beneficially readjust their lives. The less I am able to enter a non-dual state, the more I am simply using my own cognitive resources to affect someone else which, for some reason, shows up as the presence of energy and focus.

For both of the examples given above, self-ratings of nonduality were only five out of ten, which were among my lowest ratings, whereas the differences in energy were either four or five out of a possible total of five. I did a lot of 'work' in my imagination in those two cases with participants ending up energized, happy and, apparently, beneficially affected. At the other end of the spectrum, with non-duality scores of eight or nine and participant energy difference scores of minus three or minus four, participants simply reported being fatigued, in some cases unexpectedly so, and, sometimes, reported experiencing less chronic pain than expected. None of this proves anything. It does suggest, however, that both effortful cognitive processing as well as altered states of non-duality could contribute to remote healing.

There are several other factors that could play a role in remote influencing. If everything has a morphic field, then ME has a morphic field. And that morphic field contains rules for radical transformation. We can align ourselves with a morphic field by imagining being part of whatever community or activities are governed by that morphic field. If we align ourselves with the ME morphic field and carry out the rituals that fall within its purview, then we could increase the likelihood of experiencing the expected outcomes. As Bartlett has said: 'With enough individuals sustaining the consciousness of our special case reality, such as Matrix Energetics, it functions consistently and reliably'.[21] In other words, I do not need to be able to do anything other than align myself with the ME morphic field which then structures the experiences of myself and those with whom I am interacting.

Still, when I am doing remote healing, everything takes place within my imagination, and people imagine all sorts of things all the time with no apparent discernible physical effects. Why should there be a difference in my case? I think that this could have to do with the commitment with which I imagine things. When I imagine a sequence of actions during a remote healing session, I pretend that it has the same reality as a sequence of actions carried out within physical manifestation. Bartlett has emphasized the significance of such commitment in a case in which he 'became' the Brazilian healer John of God. Bartlett says that he looked at a picture of John of God, aligned himself with John of God's morphic field and in his 'mind's eye' he 'inserted John's surgical clamp right up the man's nose and into his brain' just as he had seen in a video of John of God. 'Note, I didn't say that I imagined it; that would not be enough.'[15] And having read Bartlett's description, I have 'resonated' with whatever it is that John of God did and done the same thing. And I do not even know what John of God's 'surgical clamp' looks like. The point is that treating what is in the imagination decisively as though it were reality could be a key ingredient in successful remote healing.

In the end, for such remote healing to work, reality must be structured in such a way as to allow it to work. In our conventionally scientific ways of understanding reality, the influence of morphic fields and wholehearted imaginary events cannot have any effects on physical manifestation. But if reality is structured in such a way that remote healing occurs, then maybe morphic fields and such imaginary events are important. In other words, it is possible that reality is structured in such a way that making changes in the imagination can directly change physical manifestation.[31]

So, where does this leave us? Well, remote influencing can occur, at least in the form of changes to self-reported energy levels as I found in my experiment. The more tantalizing question is whether ME or similar sorts of strategies can effect clinically significant outcomes. To answer that question would require a different research design. I encourage others to take up that task.

In general, I sometimes like to distinguish between input and output to the psyche. Precognitive dreams are an example of anomalous input whereas remote healing is an example of anomalous output. Anomalous input seems to occur more readily and be easier to detect than anomalous output. I was convinced that I was receiving information through anomalous means long before I could establish that remote influencing really was occurring in some cases. And the two seem intertwined, as we can see from some of the examples of remote healing.

I had not intended to get as involved in remote healing as I did, nor did I expect that the healing that I did would make a difference for the participants in my experiments. I also did not expect that doctors would find a lesion in my liver that would put my dreaming and healing skills to the test. Well, I did see beforehand in my dreams that something dramatic was going to happen. So here is the strange story of the liver.

Chapter 4

Dreaming Through a Health Crisis

Most of the students in my classes are excited by the course material and eager to learn as much as they can about anomalous phenomena that occur in the context of altered states of consciousness. But there are skeptics. And I encourage skepticism. A failure to be sufficiently skeptical can lead to the sort of loss of self-determination that is characteristic of cults. I keep telling students to follow the empirical evidence wherever it leads them. But some students are pathological skeptics in that they refuse to acknowledge any evidence that conflicts with their beliefs about the nature of consciousness and reality, irrespective of the quality of that evidence. This is usually most evident with students with materialist beliefs, but can also occur for those aligned with some fundamentalist religions. At that point I try to appeal to students' common sense. In the case of dreams, for instance, there is a practical matter. Dreams can help us to negotiate difficult situations in our lives and if we refuse to acknowledge that dreams could mean anything, we cut ourselves off from a valuable resource that is available to us when we could most make use of it. I am reluctant to talk about matters that affect my own health, because of the privacy of such information, but have chosen to do so in this chapter in the hope that what I say will help others to find ways to use their own resources for helping themselves.

While I was at the Asia Consciousness Festival in Hong Kong several years ago a colleague from Toronto who was also at the festival told me that I should talk to Jeanette Wayne, a holistic healer in Toronto, because I would be interested in the altered states of consciousness that she experienced when she heals people. When I got back from Hong Kong I contacted Jeanette and set up an appointment to interview her. However, Toronto is a two-hour drive from where I live and I was too ill with flu-like symptoms to venture out that far, so I called her on the telephone instead. She asked me if I wanted her to heal me. I said sure. She asked me what my main complaints were. Then told me to change my diet and to get some specific Chinese patent remedies. Then she hung up the phone and worked on me remotely for a half hour. I did not notice much during the time that she was healing me except for some activation of energy at the top of my head and the perception of a supernova contracting. I called her back after the half hour and she asked me if my knees were lighter. She said that she had seen a dense energy inside my knees.

I had not said anything about my knees, but I have had osteoarthritis in both knees for many years, probably the result of working as a roofer for 5 years and about an equal number of years practicing the martial arts. My left knee had acted up to the point where I could not skate for a while to play ice hockey. An ultrasound had shown damage to the cartilage and three cysts at the back of the knee. Eventually I had decided to play anyway, and the left knee had been holding up reasonably well. By coincidence, I was scheduled for magnetic resonance imaging (MRI) of the left knee a couple of days after Jeanette's session to determine how badly the cartilage was damaged. If the situation were not too bad, then injections of hyaluronic acid could be possible to restore function to the knee. I went for the diagnostic procedure and was surprised by the results. The cartilage was normal and there was only one cyst, smaller than one centimeter in size, at the back of my left knee. Hyaluronic acid injections were not necessary.

The other unusual thing that happened was that I had a deep and refreshing sleep the night after my session with Jeanette. In fact, I slept better than I had slept in years for several nights following her treatment, although that dropped off back to normal within about a week. This is similar to what I found with the participants in my remote healing studies. As in one of the examples that I gave, sometimes participants slept unusually deeply. Such deep sleep is conducive to restoration of the body.[6] Even without knowing much about her, I had some reason to suppose that Jeanette could be an effective healer and so I started to drive to Toronto on a regular basis for treatments by her.

The period of time during which Jeanette was treating me coincided with my remote healing experiments and the one would inform the other. For example, one Sunday evening, while I was at the beach watching the sunset and reflecting on the nature of reality, I noticed that my left ankle was vibrating. I found this to be so unusual that I was uncertain to what to attribute it. I wondered if Jeanette were healing me. I texted and asked her and got the response: 'hehehe'. When I saw her in person at a later time, she explained that she did remote healing on Sundays. This is not necessarily evidence of anything anomalous but, when one of the participants in my remote healing experiments had said that she had become aware of a pulsating feeling in the fingers of her left hand around the time that I was imagining activities that would increase the 'energy' in her hands, I had a point of reference. It seemed as though I had been a recipient of that sort of intervention and so it was easier for me to imagine that I could also have caused that type of effect.

But this also worked the other way round. Because I could 'see' things going on with my participants, and because my imaginary interventions sometimes seemed to help them, I could understand how Jeanette could be much better at those sorts of things than I, and I could accept the things that she said and did more readily than I could, perhaps, otherwise. I recall one afternoon when I had just lain down on her treatment table. Jeanette leaned over me and announced: 'You're not going to die'. I laughed. I had not thought that my situation had been that dire. But then, that seemed to change. It started with a series of dreams about a year later, starting with this one on September 1, 2010:

> ***Waking Dream:*** *I was in a house. There was a large, dark cloud outside, that had a blob that had descended all the way down to the ground. It was coming toward the house that I was in. There was a handful of people walking in front of the cloud. When the cloud reached my house, it came in through a door. I was undismayed, but directed my thought toward it to dissipate it. This didn't seem to work at first, but I persisted, and continued to pierce the cloud with my mind. I have the impression now that it seemed to swirl and break up a bit. I'm not sure what eventually happened to it.*

There's a blob in a dark cloud coming toward me. And the following night, September 2, 2010:

> ***Dreams:*** *I was running barefoot fast wearing fairly skimpy swim trunks partway around an island. At one point I looked down to see if I had been damaging my feet and saw that they were fine. I realized that there is going to be some discomfort as one exercises one's abilities in order to maintain them at their optimal levels; in this case running barefoot.*

Neither of those dreams seemed significant to me at the time, but the following dream, a few nights later, on September 5, 2010, brought me up short:

> ***Dreams:*** *A young woman was standing in front of me. She was physically fit and looked healthy, but when I looked at her tongue, there was a yellow, somewhat crooked band down the middle of her tongue. There was a voice-over of some sort saying that she needed carnitine.*

Jeanette practices traditional Chinese medicine and she often asks to see my tongue which can presumably reveal things going on inside the body. I interpreted the young woman in this dream as myself. She looks ok from the outside, but there is a problem on the inside as revealed by the band down the middle of her tongue that looked a bit like a lightning bolt. This suggested to me that there was something wrong with my body that I could not see from the outside.

The voice-over was interesting. The dream architect seemed to be suggesting that I needed carnitine. When I checked the Office of Dietary Supplements of the National Institutes of Health website,[32] I found out that, whereas people on conventional diets consume about 60 to 180 milligrams of carnitine a day, vegans only get about 10 to 12 milligrams. The body does synthesize carnitine from lycine and methionine, but it was possible that the levels of these two amino acids were also low in my body. At any rate, I thought it would be prudent to supplement with a little carnitine, so I asked my physician for a prescription several days later when I saw her for my annual physical.

Then, on September 12, 2010:

Waking Dream: *I was trying to get to a talk that I really wanted to hear. And there were several things I was trying to get done before then. I ended up coming all the way back to my place even though I hadn't intended to, when I noticed that I had left a closet-like room unlocked. I had put the chain on, but hadn't locked it. Someone had opened the door and reached in. There was an old purse belonging to mother that someone had probably raided, but I figured that they couldn't have taken very much because there wasn't anything of value in this storage closet. I opened the door and Stacey came out. I hadn't realized that I had left her in there. But she was emaciated, limping, and parts of her fur were missing. She looked more like a squirrel than a cat. She had been in the closet for too long without* *food or water. Especially water. I wasn't sure she would live. I was so upset with myself I was wailing out loud.*

My mother represents my personality, and sometimes my physical body, in my dreams. Stacey is my cat in real life and could represent a part of myself. In the dream I am busy with my academic work, yet end up coming back to 'my place' suggesting an interruption to my work by some personal needs. In the dream, I have neglected my cat, which I interpreted as having neglected some part of myself that required nutrients with which I had not provided it. That is how I interpreted the dream at the time. It was only later that I began to interpret cats in my dreams as magical healing power. Under such an interpretation, something has gone seriously wrong with my healing ability. In any case, this dream also gets my attention. Something is not right.

The liver blob

Two days later, on September 14, 2010, during what was supposed to be an innocuous abdominal ultrasound, the technician was taking an inordinate number of images of the right side of my abdomen, then said that she had to consult with the radiologist to make sure that she had everything that she needed. Hmmm. Clearly she had found something. I wondered if it were a cyst on my liver. Or a tumor of my liver caused by extremely high levels of silver in my body, as indicated by previous analyses of my hair.

I had barely arrived home from the hospital when there was a message on my answering machine from my physician. Her voice sounded grim. I called her back. The doctors had found a 'lesion' in my liver. They did not know what it was. It could be anything. But it looked suspiciously like cancer. I asked if it could just be a cyst and she said no, because the ultrasound revealed blood flow to its periphery. My dad had died of liver cancer, so I automatically assumed the worst. My physician was going on about a referral to a surgeon and how they might be able to just cut out the right lobe of the liver. At some point I was not really listening any more. It was all too horrific to hear.

About an hour later, I wrote in my diary:

Well. I'm taking this surprisingly well. But I feel as though my life has been permanently altered with today's news. I'm not the same person now as I was a few hours ago. And I don't see how I could ever go back to a state in which I simply live in this world without consideration for what comes afterwards. I need my dreams to guide me out of this. My guidance. How do I negotiate this state of being?

I lay out in the sun for a while, then got in my car and drove out to the beach. Ok. We did not actually know that it was cancer, so within a few days I stopped using the 'c-word' and called it a 'tumor'. Then I stopped using the 't-word' and just used the expression that the physicians were using, i.e., 'lesion'. Then, eventually, I just called it a 'blob'. That sounded friendlier. And it is just now, as I was writing up the sequence of dreams leading up to the ultrasound, that I noticed the word 'blob' in the description of the first dream, a blob inside a dark cloud. That certainly fit. I was definitely in a dark cloud, and the liver blob was very much a part of that cloud.

The other thing I did was to ask the blob what it was doing there. In my imagination, of course. In my imagination I received an answer: 'To create a sense of urgency'. Yes. I could see that. I felt that I had been drifting a bit in terms of my resolve to understand the nature of reality. A crisis such as this certainly sharpens the focus. I acknowledged the assistance of the blob. Now that I had gotten the point, I felt that it could just leave.

I should clarify that what I am doing here is describing my own reaction to this situation, and that this narrative is not meant as a recommendation for anyone else. I feel that each person needs to make her own decisions in light of her unique circumstances and the best professional advice that is available to her. What I am describing here are the decisions that I chose to make at the time. I would not necessarily make the same ones were I to find myself in a similar situation again.

I was supposed to go for computerized axial tomography imaging with contrast dyes but decided to turn that down as being too toxic. If my body were going to stand a chance with whatever this was, I believed that it needed all the help that it could get. That meant as much as possible staying away from what I perceived to be toxins. I asked for an MRI instead. And for that I had to wait six weeks. That was fine with me, because I felt that it would give Jeanette a chance to get whatever this was under control.

I drove to Toronto and showed up at Jeanette's clinic unannounced. I started blubbering something about a tumor in my liver when she asked 'Whose is it?' Her question did not register as I went on about how my dad had died of cancer, and so on. She stuffed me into a treatment room, fed me some tea, and looked in on me from time to time as she attended to her other patients. When they were all gone for the day, we sat in the kitchen that doubled as her waiting room.

As we were sitting there, Jeanette said: 'It's not your cancer. You picked it up two to four months ago when you healed someone'. I said 'You think that I picked this up?' And she said 'I don't think; I know. Ninety-nine point nine nine per cent of the time when I say something like that, it is correct'. I started talking about someone I had been remote healing who had died of cancer the morning of my ultrasound. 'No, it's not her'. Well, 4 months previously, I had bumped into a friend of mine in the morning. That friend had said that she had been up all night because one of her relatives had been admitted to the hospital. The doctors suspected massive internal bleeding and the relatives feared that she might be dead by the morning. She had had throat cancer several years previously for which she had been successfully treated. I told my friend that if she could get explicit permission from her relative, I would do remote healing for her that evening.

In the afternoon, when I started thinking about my friend's relative, I thought that I could end up helping her to cross over as I had apparently done for someone a week or so previously in the context of my remote healing experiments. As I was thinking along these lines, I got a bad nosebleed. In fact, it was so bad that I had to stuff cotton into my left nostril, something that I could not remember having ever had to do before, and something that I have not had to do since. I wondered if I were 'tracking' my friend's relative's condition. 'Tracking' refers to displaying in my body what is going on for someone else in her body. Usually it consists of feeling sensations in various parts of my body corresponding to parts of the other person's body where there could be some disequilibrium. Bleeding was a little extreme.

I did remote healing for her in the evening. When I 'looked' at her, she looked cheerful. I thought that must be a misperception, because I could not imagine anyone being happy under such duress in a hospital. However, as I found out later, that 'perception' turned out to be correct. I decided to continue on from the tracking theme and imagined myself bleeding for her; deliberately taking on her illness and then eliminating it from my body. This is not an ME technique but something that I thought that I would do anyway. It seemed to me that a person could absorb whatever is disagreeable into herself and then transmute it into something beneficial. The trick is to remain steady during the transformation process so as not to be affected by the disagreeable stuff. I also did the *not there* technique whereby I imagined that whatever problems she had were not there. At the end, I saw torrents of water gushing out of a pipe suggesting that whatever was wrong with her was being flushed out. When I 'looked' at her again, she 'looked' normal. This took 23 minutes.

I did not think anything more of this. I found out later that on the following day, all the diagnostic tests that were done that day had come back normal, and the morning after that, she had gone home. When I asked about her a half-year later, she was still fine. It was never determined what had been wrong with her or how serious her situation had been.

Jeanette said that that was the one. That was the incident that had resulted in my malaise. I said that remote healing does not work that way, that it does not harm the practitioner. Jeanette said that if someone is committed to healing, then healing will take place, and it will follow the path of least resistance. In this case, the path of least resistance was through my body. And now my body was trying to eliminate the other person's disease. That is why it was in the middle of my liver; because it is the liver that eliminates things from the body.

Jeanette said that when she first started treating people who were in the end stages of cancer, she did not know what else to do to heal them, so she would take the cancer into her own body. She said that at one point she was walking around with 6 to 8 tumors each of which was supposed to be lethal within weeks. What she did not tell me at the time, but told me much later, was that the cancers started to spread in her body. And that she had to learn ways to halt and reverse that process. She said that it took her 6 months to get rid of the tumors. That she had not thought that it would take her so long to do that. But that she did not regret having taken them on. Several months later, when I asked Jeanette's mother about it, her mother said that she remembers that period in Jeanette's life. That the tumors were visible underneath her clothing and that Jeanette looked emaciated and gravely ill.

Well, I seemed to follow Jeanette's trajectory. I lost 17 pounds, looked awful, and a blood test for carcinoembryonic antigen (CEA), apparently used for tracking some cancerous tumors, came back slightly elevated. My physician became convinced that

I was dying of cancer. I was afraid that I was dying of cancer. Of course, the austere diet of a tiny bit of rice every day with cooked leafy green vegetables that Jeanette had put me on could account for the weight loss. The stress of thinking that I was dying of cancer would explain my gaunt appearance. And who knows what the slightly elevated CEA level was about.

I went back in my diary to the time that I had done the remote healing session for my friend's relative that was supposed to be the source of my distress. Here is a dream from the week before, on May 6, 2010, which I found just now as I was writing this:

> ***Dreams:*** *I was lying on my right side in my dreams when a cat came up to me and started to bite my chest. I woke up with a panic attack, heart pounding, afraid that I had had a heart attack.*

Given that cats were sometimes associated for me with the magical properties of remote healing, the cat biting me in my dream suggests that I was being adversely affected by the same magical properties. Perhaps. I did not predict this at the time of the dream, so I do not want to place much emphasis on it. One can read pretty much anything into anything, so for me it is important to make predictions at the time of a dream and then watch to see if the predictions are correct. Nonetheless, this dream could well have been precognitive.

A few days after that remote healing session, when I had gone to see Jeanette, I had started to tell her about it when she interrupted me and said 'cancer'. Then I went on to tell Jeanette the story. She had not said anything further, except to tell me that there were some Chinese herbs that I needed to take. It turned out they were for detoxification. Jeanette was clearly concerned already at that time about the consequences of my remote healing stunt.

One week after the ultrasound, on September 20, 2010 I had the following dream:

Dreams: *I was going into a building for some sort of performance along with a whole bunch of other people. I was carried along a conveyor belt to the other end of the building. Then I had to go through a tiny square hole and twist myself around some corners. Like a tiny tunnel. I didn't think my head would fit the rectangular space of the tunnel, but I trusted that if I were in the tunnel then I could navigate it, so I stuck my head in and it managed to fit.*

Here is the commentary that I wrote at the time:

The dream indicates that I feel as though I am navigating close quarters in following the path of healing from the tumor. In the dream I trust that no matter what it looks like, I can manage to navigate the course that I need to take. I did think last night as I went to bed, that I can rely on the guidance that I get for getting me through this process. This dream reassures that there is a path and, however narrow it might get, it is negotiable.

This is what I had asked for. I wanted my dreams to show me the way through this crisis and with this dream, they seemed to be telling me that they would do so. My dreams were showing me that it might not look as though I can even handle the life path that I was on, but that, narrow as it might be at that time, I could manage it. And, indeed, I was managing it for all that it seemed as though it were not possible to do so.

A waking dream in the middle of the night of September 29, 2010:

Waking Dream (in the middle of the night): *I was in a room that was getting warmer and warmer when I realized that there was a bomb in the room that would go off when the room reached a specific temperature. I immediately turned down the thermostat which set the air conditioner going, but I didn't care about that. Others found the bomb and defused it and the danger was averted.*

I was not certain to what the heat in the room referred, but it could simply have been a reference to stress. I thought that the bomb in the dream was the liver blob in real life. The blob was fine as long as I did not allow my anxiety levels to rise. Turning down the thermostat corresponded to turning down fear. In practice, I continuously monitored my thoughts to dispel worrisome rumination as much as possible when it arose, as a way of lowering anxiety.

I thought that the heat could also be a representation of my efforts to mentally 'push' the liver blob into non-existence. Such 'pushing' could be counter-productive, reminiscent of Richard Bartlett's admonition not to engage at the problem level. Under what conditions *does* reality change? It changes when there is 'resonance'. So I tried to create more resonance in my life. Again, through mental vigilance. But also by driving out to the beach and appreciating the beauty of nature. Sitting with my back against a pine tree. Meditating. Those sorts of things. The dream ends optimistically with others finding and defusing the bomb.

The next significant event occurred when Angie Aristone, a medium, came to my class as a guest lecturer. I had told her about the blob before the class. During the class, she was demonstrating what it is that she does, when she was drawn to a student to her right and one to her left. I told her to go with the student to the left. She went with the student to the right and, among other things, correctly told her that her grandmother used to make her own brooms. Finally when she was through with her, she turned to the student to the left. She appeared to have contacted the girl's great grandmother who had lived into her nineties and eventually died of old age. Angie told the student that her great grandmother was telling her that there had been many lessons that she had learned from her. One of them was that you do not have to die, that you can continue living if you want to; that you determine when you die. At that point Angie turned to me and said 'And that is for you too!' Ok! That really hit me.

What hit me was the implied efficacy of our intentions. We usually think that the physical things in the world around us, including our bodies, function autonomously from our minds, and that we have little control over what they do. In particular, disease is the result of genetic dispositions, various stressors, environmental contaminants, and so on that we cannot ultimately control. What my student's great grandmother appeared to be saying is that we have the ability to override any such circumstantial factors when it comes to the health of our bodies to the point where we can live as long as we want.

I was reminded of research that provides some support for that notion. Years ago, when I taught developmental psychology, I found it interesting that *cognitive decline,* whereby an elderly person loses cognitive functioning, is a significant predictor of death irrespective of a person's actual physical condition.[33,34] Given that what is happening in our minds is a better predictor of death than physical deterioration, this suggests that how and what we think can affect our longevity. Is it possible that someone could live as long as she wanted to live simply by setting the intention? I started to question the extent to which I was a victim of my body and tried to shape my destiny in spite of how dire the situation might appear.

The situation eases

A dream on the night of October 23, 2010:

> ***Waking Dream:*** *I was with a group of people in the evening. We had made a recording of a collection of songs. And someone started to sporadically play them. I had a song in that collection and it was one of the first to be played. I had sung it well and it seemed to be a bit of a solo song with my voice rising over the rest of the music. We were getting emotional about the tenderness and compassion or whatever it was that was displayed in the songs. I thought that we*

should have just planned on playing all of them one after the other as we walked. Then we were walking back from where we were to the main road. It had the sense of walking back to normality. It was dark out. Everyone else was walking in bunches. I was walking alone between the bunches, but didn't mind the space to myself. I noticed that there was a banquet hall over on the right where people could celebrate occasions. And one on the other side of some defining boundary over to the left, but that one was seedier. I woke up.

When I interpreted the dream, it seemed to me that the people represented others who were experiencing a serious health crisis. I found that tenderness and compassion were emotions that had been released during this time, and these were being expressed in the dream. But we were walking back to normality where everyone else walks. I was walking by myself. And walking in the dark; that is to say, without knowing that I was walking back to normality. This dream suggested that my life is slowly returning to normal.

I went for the MRI. This time, there was no terse message on my answering machine. I did have a sequence of dreams a couple of nights afterwards, though, on October 31, 2010, while I was waiting to hear the results of the MRI.

Dreams: *I put through the paperwork to give away a monkey. It had been an experimental monkey, but instead of killing it, it was being given away to some lucky person at main campus at the university. The administrators checked the paperwork and everything was in order. They knew it was ok to give away the monkey.*

Dreams (cont.): *Someone asked me point blank how long I was going to live. I replied that I thought I was going to live for a long time.*

Dreams (cont.): *I was feeding dinner to another guy. I had ground* *up the body of a dead guy and put it in the soup that I was feeding both of us. The soup was actually a swamp out in a natural setting. I knew that I had done a good job of transforming the body so that the soup was safe to eat, but my companion was suspicious about it. Then, as he took a spoonful, he fished out an empty glass bottle. Ok. So I had not quite managed to get rid of everything, but it wasn't even anything resembling what had gone in; it was only a glass bottle that looked like an empty President's Choice Organics maple syrup bottle with the cap on. I didn't care.*

All three dreams have the same message. I've got the monkey off my back in the first one, I say that I am going to live for a long time in the second one, and whatever the blob is, it is inert. The third of those dreams was particularly significant to me. The 'body of a dead guy' signified whatever it was that had been wrong with me. If I had picked up someone's cancer, then that cancer was gone. The 'spoonful' is the MRI image. It reveals that all that is left is a container. A glass container that cannot contaminate the swamp that it is in. The swamp represents the liver. There is nothing toxic left. The healing has been successful.

I thought this dream was showing me what was happening. I had hoped that whatever was in my liver had completely resolved itself so that the MRI would come back normal. Not so, according to this dream. There was still something there, although it was less frightening than what it had looked like previously. Then, on the night of November 4, 2010, another dream that seemed particularly significant to me:

Dreams: *I was driving and there were French doors that were slightly closed so that the opening through which I had to drive was not quite big enough. I realized that I needed to get out and open the doors so that I could get my car through.*

Yes, not quite over. French doors in the way. Actually, not even

French doors, really, but a pair of doors that swung toward me. I interpreted this dream to mean that there were still two things holding me back. The good thing was that in the dream it appeared as though I could open the doors. That suggested to me that the barriers could be removed so that I could continue to drive.

Several days later I got a phone call from my physician. The MRI revealed an irregularly contoured 'lesion' in the liver that measured 5 x 3.9 x 4 centimeters. The size on the original ultrasound had been 5.4 x 4.3 x 4.9 centimeters. So, whatever it was, had apparently not gotten any bigger, and looked as though it might have gotten smaller. But the doctors still did not know what it was. They wanted to use contrast dyes or agents to further define it. I was at the low point with my weight. My physician thought that this could be cancer that had metastasized from the colon. I was still supposed to be going for a consultation to a surgeon. This was reminiscent of the dream of October 31, 2010 in which I felt that everything has been resolved, but my dream companion was distraught.

At any rate, apparently I had not yet drained the full value of the sense of urgency that the liver blob had been instilling in me. But there were two further tests that could help to restore a sense of normality. One was a blood test, the alpha fetoprotein (AFP) tumor marker, used for tracking some tumors of the liver, and the other was a colonoscopy. These were the two doors in my dream.

Before getting the results of those two tests, I had another dream, this one on November 25, 2010:

> ***Waking Dream:*** *I was helping out some guys doing construction putting in some ventilation pipes in the ceiling of a building. It was pretty dark in there. I thought that we needed the square covers for them that ran sideways along the length of the pipes, and then realized that the stuff that was sitting there packaged contained*

those parts and all we had to do was install them. There were some old notes in brief cases that other professors had left behind years ago in a back room and no one could throw them out because they needed permission from those professors to do it. But the professors were long gone. I figured that if I were in charge, I would just throw out all the old notes.

The building is almost ready. That could represent the idea that some new form of expression for myself is ready. The brief cases are interesting. In the dream, the cleaning staff keep putting the brief cases back, even though the professors who placed them there had been long gone as these had been courses that had been taught long ago. These could represent inactive tumors that have been deposited in various places in my body resulting from some lesson that I needed to learn at the time. In my dream, there was a briefcase that looked like the tan briefcase that I used to carry in my first year of university, when I was an Engineering Science student at the University of Toronto.

When I thought back to my first year of university for something relevant to my health, I recalled that I had taken high doses of a tetracycline antibiotic for about a year or more for my acne. Then, at one point in time, I ended up suffering from a sore throat and a high fever that would not come down. I was hospitalized for about ten days while I got over whatever infection I had had. This was a turning point in my relationship with the medical profession. I had trusted my physicians and dutifully done whatever I had been told to do. I had not cared that I had had acne and it had not bothered me, but I had been told to take pills for it so I had taken pills. What I learned was that the pills could have suppressed my immune system and made me vulnerable to viral infections. The medical staff at the hospital never did identify the pathogens that had caused my sore throat. It took me a while to recover, and when I did, I kept my distance from the medical profession for several decades and, instead, explored the services of alternative health care providers. I also assumed responsibility for my own health and did not allow myself to become involved in anything with which I did not feel comfortable, such as being injected with contrast dyes and agents for diagnostic imaging.

I told Jeanette my briefcase dream and my reflections about the tetracycline. She had an *aha* moment and said that that was a source of weakness in my body. I lay down on her treatment table. This is the description from the write-up in my diary:

> *Jeanette didn't stick any needles into me, just put her right hand on the right side of my abdomen in the area of the liver, and her left hand underneath me against my back, opposite her right hand. After about 5 minutes of this, I felt a gentle vibration in my body under her right hand that lasted for about a minute. Then my abdomen just collapsed inward about 1 to 1½ inches. It was an odd feeling. It just felt as though something disappeared.*

Well, this was unexpected. When I told Jeanette what I had felt, she said that it is nice when her patients can feel what she is trying to do. When I asked about it again several months later, she said that a pattern of information had left. That usually those just leave, but that sometimes a person can feel them leaving if there is an educational purpose for the person knowing about it.

However, my dreams from the night of December 5, 2010, were anxiety arousing:

> ***Dreams:*** *Again, there was something wrong with me in my dreams.*

> ***Waking Dream:*** *My car was smashed. I was looking down on it. Then my car turned into me. I woke up.*

According to the dreams, something is wrong with me. And, in real life, I do not feel healthy. I feel sick. I worry that I am sick.

So at that level, these dreams are accurate. But they are inconsistent with the previous dreams showing me that whatever was happening in my liver was not malicious. I have included them to indicate the need to retain a dispassionate attitude toward the dream content and to continually seek to discern the meanings of dreams.

Let me be more specific about the need for discernment. Many of my dreams are simply reflections of my fears and desires and do not accurately portray the reality within or without me. I am afraid of the colonoscopy and the blood tests. I am afraid that they will reveal that I am seriously ill. That fear projects into my dreams. I can usually recognize such dreams immediately and see them as the reflections of the fears that they are. However, the second dream is a waking dream and, as I said previously, that space in my mind as I wake up is *sacred space*. I interpret that as the space in which the dream architect can impart wisdom that stems from deeper parts of myself giving me insight into what is going on in reality. It is usually representational. Unlike the first dream in which I just think that there is something wrong with me, in the second dream, my car symbolically stands in for me, at least at the beginning of the dream.

There is also the possibility that I was in a state of superposition in that both realities were true at once: I was seriously ill and I was essentially healthy. It could be that it was up to me to choose which alternative I wished to actualize. That would be consistent with my student's great grandmother who said that we get to choose when we die. Perhaps the choice of whether I was ill or well was up to me.

It was time to do some healing for myself. I realized that I had spent enough of my life anticipating the worst. I used every technique in the book. Two-point. Alien head. I brought in something from behind a window. I ran a cycle of affirmations that I had been using to restore health. Alternate selves. Time travel. I tried to optimize the use of my time to change myself while I did not yet know the outcome of the medical tests. There was a snow storm outside as I was doing the healing.

Let me just elaborate on the statement that 'I tried to optimize the use of my time to change myself while I did not yet know the outcome of the medical tests'. I try to treat every situation in which I find myself as an opportunity to learn something and to develop myself into a better person. So it is with difficulties such as health crises. The tendency is to want to escape from difficulties and to try to get back to 'normal' as quickly as possible. I felt that some version of 'normal' was coming, but I was not certain of that, so I was still significantly stressed. What I was trying to do was to extract all of the learning potential from the stressful situation before the stress eased.

The night before the colonoscopy, on December 7, 2010, I had the following dream:

> ***Dreams:*** *I was somewhere with a friend and her kids. One of her huge dogs was there. He had one of my forearms firmly in his mouth in such a way that I could feel the impressions of his teeth and I could not get my wrist out. I wasn't too afraid that he would bite it, even though I realized that he could easily do that. Then my wrist was released. And there was a young family there of mixed race who had a really sweet little girl. She was walking around but had just been born recently. I realized that my friend might not know her because her kids were a bit older and so would not have played with her.*

Ok. I would be ok. I could not avoid the medical tests and the results could be disastrous but, according to the dream, they would be ok. Having my forearm trapped in the dog's mouth represented the medical tests. The possibility that the dog could bite me signified that it could turn out that I was seriously ill. But in the dream, the dog did not bite me and released my forearm, suggesting that the tests would not reveal anything malicious.

And there was a symbol of rebirth in the dream; a new me afterwards. Just as the previous car crash signified the loss of my 'vehicle', the 'little girl' who had 'just been born recently' represented a new personality structure. And indeed, the colonoscopy and the AFP tumor marker were both normal. I had opened the French doors and could continue to drive.

Variations on a theme

There was something that the endoscopist said to me before the colonoscopy that stayed with me. He said that the ultrasound and magnetic resonance images could just be 'shadows'. In other words, they could be artifacts of the imaging process. That was one interesting variation on what was going on. Here is another.

I was reading through the proceedings of a conference about carnitine, when I came across the following information: Carnitine was being used 'in the treatment of patients with nonalcoholic steatohepatitis. . . or fat deposits in the liver [which] can result from . . . the antibiotic tetracycline'.[35] Now this was interesting. Apparently tetracycline can leave fat deposits in the liver and carnitine is being used to treat that condition. I asked my physician if my liver blob could just be a big blob of fat, given that I had taken so much tetracyline earlier in my life and that I had had the dream suggesting that I take carnitine. She said that fat usually gets spread around and does not just accumulate into a single blob. However, it is possible that the carnitine that I had been taking had counteracted the 'weakness' that Jeanette had perceived by metabolizing whatever fat deposits had still been left in the liver, and that my liver was 'stronger' as a result; better able to deal with whatever else was happening to it. And that Jeanette made whatever residual effects were present, disappear, although I do not know what it was that palpably disappeared.

I had a dream on the night of January 3, 2011:

> **Dreams:** *At one point I looked through a lens with my eye close to it and could see my mother. Then, when I tried to look at her afterwards through the lens, I couldn't get the aim right. I also realized that my eye had to be up close to the lens, otherwise I couldn't see in focus. But then when I got my eye up close, I couldn't get the aim right. At that point I realized that the best glimpse I had had had been that one time earlier.*

Given that my 'mother' typically represents my physical body or my personality in my dreams, I interpreted this as a dream to see what was happening with my body. So, I get a close look at my body by looking through an amplification device. But then, afterwards, I just cannot seem to get the picture either in focus or on target, and I realize that the best look I had was that one time. The conclusion I came to from this dream was that the best image of what was going on in my liver would come from the MRI that had already been taken. That further imaging efforts would be less informative than the earlier one.

I did indeed go for another ultrasound. The technician asked me if I had had an impact to the liver. As a matter of fact, I had been struck hard, a number of times just below the rib cage less than two weeks before the first ultrasound. I had been in pain for several hours after this occurred. A few days later, I had the dream of September 5, 2010, showing me that there was something wrong with me. When I spoke to the radiologist about it, she acknowledged that this could be a hematoma with scar tissue, that is to say, a bruise to the liver. When I talked to my physician about it, she said that the aftereffects of a bruise could persist. That the blood that had diffused in the liver could remain there permanently and that that could become a cyst with an outer shell and a hollow interior. This time the blob measured in at 4.6 x 3.9 x 4.3 centimeters. Not bigger than previously. Time to stop worrying.

I did not think much about the blob, but perhaps there were

still something going on with it. I had the following dream on April 24, 2011:

> ***Dreams:*** *I was driving Jen with passengers in the car along a side road and there was snow on it! I had already put the summer tires on. In fact, I had to drive sideways at one point and thought I would fall into a river, but we made it ok. Later on in my dreams again, there was snow.*

This dream could be relevant. Jen is the secret name for my new car. The fact that I have put on my summer tires yet there is still snow on the ground could signify that I think that my troubles are behind me, yet some still lie ahead. This is echoed in a dream on May 8, 2011:

> ***Dream:*** *I'm walking along. There are other people around and I'm about to turn right up a hill. It's natural countryside. But there's a lion to my left lying on the ground. I try not to be afraid, turn my back, and head up the hill. As I head up the hill the lion follows me. Now it is over to my right side, but much smaller, about the size of a lynx. I am no longer as afraid of it. I stop and now it is a strange creature with a long snout; soft like a ferret or something of that sort. It says 'Please don't shoot me'. I put my arm around its head and put my head up against its head. I feel pure love for it. I think I reassure it that I will not shoot it.*

What is going on? Ok. The lion could be my personality and this is an allegory about my relationship with my personality. Or the lion could be the liver blob. Or both. I find that often concurrent interpretations run on the same imagery, as though there were a pattern in my reality playing itself out at different levels. In the dream, I try not to be afraid. I turn my back. In the interpretation, I forget about the liver blob. But it follows me. I am less afraid of it. And it changes. It is smaller. It becomes a creature with a snout and asks me not to shoot it. In the dream there is a transition from my being afraid of this creature to the creature being afraid of me. I do not know what that is about.

Another ultrasound. This time the blob appears to be bigger, at 5.4 x 4.4 x 5.5 centimeters. That would be the snow on the ground. That would be the creature showing up again after I have turned my back on it. My physician thinks that the variations in size are just due to measurement error and that whatever it is that is in my liver has not changed substantially. Nor is it interfering with liver function as determined by blood tests. She thinks that the blob could be a hemangioma, which is apparently an aberrant complex of blood vessels. In fact, on the night of July 14, 2011:

> ***Dreams:*** *I was looking at my body (or some body) and there was a warren or network of some sort in the body, lower down, that was not quite full size, whatever that size was supposed to be.*

I realized later that there was something to do with the head as well. Perhaps a much smaller warren in the head. Or maybe this was just a 'head trip'. One of the things I have noticed is that sometimes dreams will lie to me in order to get me to do certain things. Of course, a skeptical interpretation would be that dreams are just junk anyway, so getting something wrong is hardly surprising. But my experience has been that sometimes the 'lies' are just what I need at the time, even if they are not true.

The warren in the dream could be a haemangioma in real life. So I browsed that flawed oracle, the Internet, and found that haemangiomas in the liver are very common with up to one out of five people having one, most of them never finding out about them. They can be diagnosed using an MRI without contrast dyes if the T1 signal is hypointensive and the T2 signal is hyperintensive. I did not know what T1 and T2 signals were, but when I dug out the MRI report, I read that the T1 signal had been hypointensive whereas the T2 signal had been hyperintensive. Voila. So I walked around for a few days thinking that there was a haemangioma in my liver.

I had bundled up all of the information about the liver blob, including a DVD with all of the ultrasound and MRI images, radiologists' reports, blood tests results, and so on, and mailed them to a liver pathologist in another city. He said that the blob was not typical of anything. And had anyone suggested doing a biopsy? Feeling some need to help out here, I was starting to think that I should perhaps agree to a biopsy. The problem is that sticking a needle into a warren of blood vessels could be dangerous. So, because the blob could be a haemangioma, it would have been imprudent to do a biopsy. And that could have been the head part in my dream. It may have just been a head trip to get me to think that the liver blob were a haemangioma so as not to consent to a biopsy.

Now for the next twist in the story. I had had several appointments with another holistic healer from another city, all but one of them through distance healing. He said that he had written down 'hamartoma' for what it is that was in my liver. He said that I had had a twin brother in utero whose fetal debris had become incorporated into my liver. I was sitting in my bedroom with the air purifier on as I was talking to him on the telephone. As he talked about the 'hamartoma', my left eye started burning and watering profusely. I felt as though acid had been splashed into my eye. I was so distracted that I was moaning into the telephone. The healer managed to calm me down so that the burning stopped, but then a few minutes later it started up again. Wow! Again he had to calm me down. In fact, the healer felt that there was bone and hair in the abnormal growth in my liver. He said that having identified the energy pattern it would be easier to remove it and that the 'hamartoma' could submit to energy healing.

I asked the liver pathologist if this could be a hamartoma. He said that hamartomas are not found in the liver. (Apparently they can be, but are very rare.) Furthermore, he said that the presence of bone and hair would technically make it a 'teratoma'. 'Hamartomas' are growths consisting of tissue that is natural to the organ in which it is found, whereas 'teratomas' can be made up of a variety of tissues. But these are usually not found in the liver either. There are some articles in the medical literature documenting the existence of liver teratomas, but they are indeed rare.[36]

If the liver blob were to be a teratoma, then the dream of October 31, 2010 seems particularly prescient. The blob would be the 'ground up. . . body of a dead guy'. The 'dead guy' would be my putative twin brother whose fetal debris I had encapsulated in my liver. There is also some resonance with the squirrelly-looking cat in the dream of September 12, 2010 and the long-snouted ferret in the dream of May 8, 2010. In both cases the creatures were in danger of being killed by myself, either through neglect or deliberate action.

How many versions of the blob are there? And why so many? A silver tumor. Cancer of the liver inherited from my father. Somebody else's cancer. Cancer that has metastasized from elsewhere in my own body. An inert shell. Artifacts of the imaging process. A blob of fat. Scar tissue from a hematoma. A hemangioma. A harmatoma. A teratoma. Does anyone know what it is? Is it even just one thing? I felt that the notion of what was going on in my liver was like an onion. Every time a layer got peeled away, another one revealed itself underneath. Is there even a single, persistent reality underneath these layers or is the liver blob quantum-like in that it is a superposition of all of them? That seems impossible, but then so did a lot of things until they actually happened to me.

Not quite there yet

It was time for another ultrasound. And a dream on the night after the ultrasound and before receiving the results, August 8, 2011:

> ***Dreams:*** *I was frustrated with medical physicians in my dreams. I was in a room downstairs in a fairly old house and some radiologist or other 'expert' showed up. A big guy with tattoos on his face. I kind of liked the way the tattoos moved and thought that he had those tattoos for the sake of children. I was ranting against medicine and was going to tell him off and that I would only deal with my own physician. He was trying to explain to me how it was important to consider the different possibilities and then to come up with the correct course of action by moving about a dozen small plastic horses forward along a surface. I was telling him that that is not actually the way that medicine works. Or science. And started to tell him about social psychology of science. I was going to mention the significant results in my remote healing study but figured that that would be too much for him to accept. But somewhere along the line he disappeared and was replaced by an older woman or man who was just some administrator or other lackey. I got angry and said that I demanded to speak to a physician. There was another lackey of some sort in the room as well. I started throwing things around and trashing the place in frustration, all the while realizing that I was just harming my own efforts since the response would be to tighten security and make it more difficult for anyone to rebel.*

I should explain that 'tell him about social psychology of science' is just short-hand for explaining the social psychological factors that mediate our interpretations of reality. For instance, we think that we are being scientific. However, we are often under the influence of the *confirmation bias* whereby we seek out information that confirms what we already believe and avoid anything that could challenge our interpretation of reality. In order to practice *authentic science*, we need to be able to follow the evidence wherever it might lead and let our theories reflect the facts.[17] So, although we might think that we have gotten things right, we could be completely off track. This is what I was trying to point out in the dream to the physician with the facial tattoos.

My dream reflected frustration with the medical profession and the complete inability to be heard above the fictitious explanation of what is supposed to be happening. And then I was not even talking to anyone from the medical profession anymore; just some administrative person. And there was no point in going on a rant because I would be regarded as a lunatic and security would just get tightened making it even more difficult for anyone to rebel. Ok. The dream was clearly about frustration. Frustration with the medical profession, although not with my family physician with whom I had good rapport. Or displaced frustration perhaps. Frustration that the situation with the liver blob had not been resolved.

And, indeed, the day after the dream, the blob appeared to be slightly larger at 5.7 x 4.5 x 5.3 centimeters, although, again, that is probably within measurement error judging by the variation in the previous sizes. I recalled that the dream of July 14, 2011 had suggested the presence of a hemangioma that could still get a bit bigger than it already was at the time of the dream, so that this increase in size could have been anticipated.

Dream incubation involves asking a question that is to be answered by one's dreams. I thought it was time to incubate a dream, so I asked the question of what else I could do to clear all of the liver problems. I got the following dream on the night of August 9, 2011:

> ***Waking Dream:*** *I was in charge of a performance in an old church or other such building. The performers were starting to assemble. I walked back to my place. As I was walking up to the entranceway, I realized that I had somewhat ok shoes on, but that I actually had a pair of shiny dress shoes that I could wear. I do not know if I put them on or not, because after that I was back at the place where we were to*

perform. Some of the performers were young adults and I realized that they knew a lot more about music than I did. Or had known, when I was their age. I was a bit senior, yet I was the one ultimately in charge. They were going about rehearsing without my participation. I had a personal helper who brought a formal-looking waiver to me that we had to sign in order to be able to use the space. I didn't want to sign anything of this sort, but I figured that I had no choice since we needed to be able to give the performance. I started reading the form and it was something about some Chinese who were asking for something to do with trading or something of that sort. I realized that it was spam. I told my personal helper to tell the guy who had printed it off, who was an old caretaker of some sort who had been there all night, that it was spam and that we could ignore it.

How did I interpret this dream? Singing in a choir or being part of a church performance have long been symbols of my service to humanity, given the dual meanings of the word 'service'. So this dream appeared to be about service that was to take place in the near future. What service? Well, telling the story of the liver blob in a book for others to read. But in what way is this an answer to the question that I had posed before going to sleep? It is an answer to my question in the sense that the liver blob is not about my liver so much as it is about my service to humanity. This incident with the liver blob illustrates how dreams can be used to track the events that occur during a health crisis. And it could be helpful for other people to see how that could play out. In other words, all that was left to do with regard to the liver problem was to talk about it.

There is also an interesting specific bit of information in the dream, namely the waiver. When I had gone for the MRI, I had been asked to accept an intravenous gadolinium contrast dye that would help the physicians to determine what the liver blob was. I had been asked to sign an informed consent sheet giving permission to use the dye. I had refused to do so, and so the dye had not been used. I was wondering now whether I should perhaps accept the gadolinium contrast dye during an MRI so that the radiologists could have a better idea of what was going on in my liver. How else was I going to be able to be of service, if I did not even know the identity of the blob? In the dream, a waiver was brought to me. I started reading it carefully and I recognized it as spam that had come to the church computer during the night. I did not need to sign it. There was no one associated with the church asking me to sign a waiver. I interpreted those aspects of the dream to mean that it was not necessary for me to agree to the gadolinium contrast dye in order to be of service. The service could proceed without it. There is even a Chinese connection, in that Jeanette, a practitioner of traditional Chinese medicine, initially tried to talk me into accepting the gadolinium contrast dye when I had the MRI.

A resolution of sorts

Almost two years after the initial ultrasound during which the blob first appeared, I went for a fifth ultrasound. On August 12, 2012, the night before finding out the results, I had the following dream:

> ***Dreams:*** *I had decided to run down a trail that had some steps in it and that went through the woods between some people's cottages. I knew that these people had dogs that could attack me but I figured that I could outrun the dogs. Sure enough as I started down the trail I could hear some dogs barking. I was running well and quickly. I made it around a right-angled corner that turned the path from going straight down the hill to going to the right, although still dropping in altitude. I was also afraid of encountering a bear in the woods. I thought that if that happened, I could stop and let the dogs and the bear have it out with each other. But then it occurred to me that the dogs and the bear could team up and both attack me. But I didn't stop.*

Unlike the earlier dream in which a dog has my arm in its jaws, in this dream I only hear the dogs barking. The immediacy of the threat has been removed. Nonetheless I am afraid of the dogs and the bear is also a symbol of my fears. But in the dream I do not encounter any dogs or bears. I just keep running. The liver blob comes in at 5.2 x 4.3 x 5.8 centimeters which is about the same size that it was previously. I just keep on living. I am the 'liver' after all.

So this story about the liver blob does not really have an ending. And that seemed to me to be an unsatisfactory ending. But then I realized that whatever ending the story will eventually have cannot change the way in which my dreams have already been helpful for me in negotiating this health crisis. That is the point of the story. If this story were to have an ending, whatever ending it would be, then that ending would be reflected back into the events that I have described. In other words, suppose that the liver blob were to be a hemangioma of the liver. Then all of the events in the story would be reinterpreted in light of the blob being a hemangioma. But those events stand on their own by drawing attention to the ways in which dreams can help a person through a stressful time of life. In this example, this was a health crisis such as that which many people face. But it could have been some other distressing series of events, such as an ongoing difficult relationship, or obstacles in one's career, or a military tour of duty.

The other thing I would like to emphasize is the importance of patience when learning to use dreams for helping oneself. Douglas Baker said that it took him 25 years before he was able to create an effective communication link through his dreams.[37] It took me about the same length of time to see that my dreams really were more than just duff. I do not think that it needs to take that long. However, part of the reason for patience is the need to develop discernment so as to be able to meaningfully interpret dreams.

The same dream images can mean different things for different people because of the associations that they have with the images. Both feared and desired images can pop up in dreams without having much meaning. Having read this chapter, readers could end up with blobs in their dreams. And, in some cases, readers will have dreamt about blobs in anticipation of reading the chapter. So having blobs show up in one's dreams would not be surprising. But how do they show up? Is the reader just processing the day's events? Are her fears or desires being reflected in her dreams? Or has the dream architect tailored the details in such a way as to tell the dreamer something? What are the unique messages that the reader's blobs are telling her? Being able to answer such questions effectively takes considerable critical thinking and practice. I am still working on doing this well. And so readers need to be patient with the process.

Finally, I want to return to one of the first dreams in this sequence, the dream of September 2, 2010, in which I was running barefoot fast in skimpy swim trunks on a desert island. I feel that now I can see the relevance of that dream. Having written this chapter, I feel exposed in the same way that running barefoot in skimpy trunks would make me feel. I feel exposed because I have revealed personal information about my health and about dreams that I feel have had a bearing on the course of my health. I realize that not all readers will be sympathetic to my interpretation of events and hence have exposed myself to criticism. I am certainly on an island in the sense that I am sitting alone in front of my computer with my mother downstairs washing the dishes and Stacey asleep somewhere, probably underneath the chair in the family room where she is hidden from sight by the chair's slipcover. Just as in the dream I feel that I need to exercise my abilities, so in waking life I feel that I need to take this step in self-exposure if what I have learned is to be of benefit to others. There is some discomfort in writing about what has occurred for me. That is unavoidable. But I am ok, just as in the dream when I looked down at my feet I realized that they were fine.

Chapter 5

Talking to Dead People

What I am doing in this book is talking about experiences that I have had that have led me to realize that reality is more interesting than we ordinarily think that it is. Such experiences have included precognitive dreams, including dreams that have helped to guide me through a health crisis, and experiments in remote healing, whereby I appear to have had discernable effects on other people during distance healing. It turns out that reality is not only more interesting than we ordinarily think that it is, but also friendlier. Not as frightening. Except that there is still that really scary end to it all: death! But is death an end to be feared? Or something else?

The scientific hypothesis that consciousness in some form survives death, at least for a time, is known as the *survival hypothesis*.[38] And there is considerable evidence that has been gathered regarding this hypothesis. I am not going to try to review that evidence here, but just discuss some of the relevant experiences and reflections that I have had. I am also not going to try to survey the beliefs about life after death that exist in various religions. I am not interested in believing things but rather knowing what there is to know.

I will begin by recounting a ghost story that was told to me by one of my students in which words appeared anomalously on a computer screen. Such events fall under the rubric of *instrumental transcommunication*, abbreviated 'ITC', whereby electronic devices are ostensibly used for communicating with the deceased. Then I will say a little bit about mediumship, in which a person apparently communicates directly with the dead. If we continue to exist after death perhaps we existed before we were born, so I will say a bit about past-life regression. Finally, I will end by talking about imagining being dead. Although we will end up discussing some weird things, there is little actual evidence for survival in what I say. However, more generally, I think that a dispassionate evaluation of the broad spectrum of available evidence tips the scales somewhat in favor of survival.[39]

A ghost story

One of my students, whom I will call 'Angela', told me the following story. Angela's mother was writing a paper for a course that she was taking. The paper required her to write about her personal life, including her beliefs, family relationships and obstacles that she had overcome in her life. While working on her references, Angela's mother looked up at the computer screen to see that the word 'perfect' had appeared on the screen. She had not written that, but did not think much about it, deleted the word and continued to work on her references. When Angela's mother looked again at the computer monitor, she saw that the same word had appeared again. This time she called Angela over and asked her if she knew what would make words appear in that manner in her document. Angela did not know what could have caused that.

It was at that point that Angela's dog moved from underneath the computer desk and startled her mother. As her mother jumped up in surprise, she hit her knee on the computer desk. Angela screamed, as she did not know what had caused her mother to jump. With that, two new words appeared on the computer screen: 'screaming' and 'ouch'. Angela and her mother noticed that the words appeared all at once rather than one letter at a time. Afraid that whoever was writing the words on the computer was attempting to hack into the system, Angela pulled out the cables connecting the computer to the Internet and unplugged the webcam.

Angela was on the telephone with her boyfriend at the time and asked him what could be causing the problem, to which he had no satisfactory reply. Angela's mother suggested that Angela's father or brother could have been playing a prank on them, given that they were the only people in the house. Angela ran upstairs and asked them if they were doing anything to the computer. Both were confused. Angela's father and brother followed her downstairs to where her mother sat at the computer.

Once Angela and her family were assembled downstairs in front of the computer another message appeared: 'get the Ouija board, or else'. Angela walked over to the laundry room where the Ouija board was kept in a cupboard. As she opened the laundry room door, her mother stated that the word 'boo' had appeared in the document. With that, Angela screamed and 'screaming' appeared next to the word 'boo'. Her family decided that they had had enough for the night and were getting ready to leave the computer, when it left one more message: 'nite'. They all went upstairs and tried to fall asleep.

The following morning, while getting ready to go to work, Angela's mother addressed her deceased grandmother, telling her that if she were to know what was going on, to tell these spirits to go away; that they were not wanted. No one was willing to venture downstairs, but Angela's mother and brother both needed to retrieve things, so they went down together. Angela's mother went over to the computer and found that there was a document open with the phrase 'sorry loves' written on it. A few weeks later Angela's mother received her paper back from her instructor. She had received a perfect grade for the paper. 'Perfect' had been the word that had started this whole sequence of events.

I have heard many such stories from my students as well as others over the years and have read many more in the survival literature so that they no longer surprise me. One interpretation of the events of this story is that Angela's deceased grandmother was manipulating the computer through remote influencing.

Another interpretation is that someone else was inadvertently influencing the computer, possibly Angela's mother or Angela herself. The question is, can phenomena such as these be produced in a laboratory? And the answer is: maybe. This is the area of investigation known as ITC. There have certainly been a number of efforts to replicate these types of phenomena although pretty much all of those have fallen outside the realm of conventional science. So I decided to try to recreate them myself.

Experiments in instrumental transcommunication

There is a historically earlier version of ITC known as EVP, or *electronic voice phenomenon,* whereby the voices of the dead can apparently be heard over the radio or on acoustic tape recordings. My first experiment was an EVP experiment in which two research assistants alternated in recording the noise from two radios tuned between stations onto cassette tapes. That was pretty much the standard EVP protocol. There were 81 sessions with an average of approximately 45 minutes per session for a total of about 60 hours and 11 minutes of recording. I had the research assistants listen carefully to the audiotapes for any anomalous sounds. Although we did get some short phrases and other noises characteristic of what other EVP researchers had heard, there was no evidence that any of what we heard was anomalous. I wrote up the negative results and had them published in an academic journal.[40]

Well, that set off a firestorm of protest. Various amateur EVP enthusiasts were outraged that I did not find anything, and I was deluged with e-mail, mail, telephone calls, cassettes, CDs, even a visit to my office, all in an effort to convert me. In their eyes I think I was just one more mainstream scientist who was so closed-minded that he was determined not to see anything if he tripped over it. Such a reaction did not reflect well on the amateur EVP research community. Science is not a matter of finding something when nothing is there. And, to my knowledge, there is no professional EVP research community. And that does not reflect well on conventional scientists who are often too biased to take these types of phenomena seriously or too frightened to jeopardize their careers by conducting research into subject areas of which their colleagues would not approve.

At any rate, I realized that the main problem with studying EVP is the difficulty of determining when acoustic sensations are anomalous given that we know from research in psychology that we creatively construct meaningful phrases when presented with ambiguous auditory stimuli. I decided to do a second study, this time using randomly printed text on a computer screen, so that there could be no ambiguity. To that end, I wrote three computer programs that generated random text: one that created random character strings from the letters of the alphabet and numbers; a second that created sentences from a bank of words; and a third that randomly just produced either the word 'yes' or 'no'. Formally, this would now be a study of ITC rather than EVP, since the mode of intended interaction was not just acoustic anymore. Oh, yes. I also got Angie, the medium who comes to my classes as a guest lecturer, to participate in the experiment. The idea was that she could talk to the dead people to see if they could give us any suggestions for getting the ITC to work.

Thus it was that Angie and I would sit in one of the rooms in the psychology laboratory, where we tried to chat up the dead people. We would also ask questions and then turn on one of the random text generators to see what output it produced. We also turned on a cassette recorder to capture any EVP either in the room in which we were located, one of the other lab rooms, or the stairwell leading out of the psychology laboratory. There were 26 experimental sessions with the text generators being engaged 715 times and producing 23,281 discrete units of textual data.

What happened? Well, Angie and I had a good time but it is not clear whether or not we actually talked to any dead people.

The first two random string generators did not produce anything decisively anomalous, but the yes/no generator did. We had asked 11 questions with unambiguous yes or no answers and 9 of those 11 answers were correct. That is a statistically rare event with a probably of occurrence by chance of only 4 per cent suggesting that something unusual had happened. But, of course, even if something unusual had happened, that says nothing about survival after death. We ourselves could have somehow affected the yes/no generator with our minds to produce the correct results.[41]

Even though the ITC research did not reveal anything convincing, by herself Angie was often able to obtain correct information about those who were deceased, or those known by them, through extrasensory means. I saw her do this on numerous occasions. At one point she decided to 'read' me and started in in her usual style: 'If I was to say January, that there's a birthday or passing connected to you in January, could you identify that now?' Well, yes, I was born in January.

The first time that I knew that something unusual was going on was during one of Angie's visits to one of my classes as a guest speaker. As part of her presentation, Angie 'read' some of the students in the class. She turned to one of the girls and said that her mother had been one of seven children in the family. That turned out to be correct.

One of the theories of mediumship is that mediums are 'cold reading' whereby they obtain cues from whomever they are reading by observing their clothing and mannerisms, sticking with high probability statements, and so on. This is certainly not something that Angie was doing deliberately, but it could be argued, was somehow doing subconsciously. However, it is not clear how someone could cold read that a student's mother was one of seven children. Nevertheless, coincidences do occur. It was what happened next, that made that hypothesis go out the window.

Once Angie was finished with the student whose mother was one of seven children in the family, she turned to another of the students in the class and, before soliciting any information from her, told her that her mother had been one of 17 children in the family. Having just lucked out at 7 with one student, if she were subconsciously cold reading, would she really be trying for 17? I mean, how many people have 17 children? But that also turned out to be correct.

I have also seen Angie correctly imitate gestures that were ostensibly made by the deceased while they were alive. For instance, in one of my classes, she told one student that his grandfather was telling him to follow his heart and not 'this', where 'this' referred to a gesture that Angie made by holding up her hands and flapping her fingers against her thumbs. I turned to Angie with a quizzical expression on my face, unable to understand the intended meaning of the gesture. The student, however, said that that was what his grandfather used to do when he was alive to indicate that he should not listen to what other people said. Angie had apparently managed to correctly imitate the gesture that used to be made by someone who was now deceased.

Talking to a dead hockey player

For a while, Angie and I got together at my place to talk to dead people to see if they had anything worthwhile to tell us. One day, the following interchange took place, as documented on a tape recorder and subsequently transcribed. The purpose of sharing this is to give some feeling for the flow of information that a medium can produce. I should alert readers to the fact that Angie is going to jump around a bit from one topic to another. A friend of mine, with whom I had played ice hockey, had recently died. This was our interchange:

Imants: *My friend has been on my mind. I used to drive him to* *hockey on Wednesday nights.*

Angie: *Is he a beer drinker?*

Imants: *Oh yeah.*

Angie: *Would you know if he drank Canadian?*

('Molson Canadian' is a type of beer.)

Imants: *No, I wouldn't know that.*

Angie: *Dark hair?*

Imants: *Yes.*

Angie: *Short dark hair?*

Imants: *Very short.*

Angie: *Lean guy?*

Imants: *Tall.*

Angie: *Yeah, like tall and lean, not a heavy set guy?*

Imants: *Yeah.*

Angie: *And you would drive and he would sit in the passenger seat?*

Imants: *Yeah.*

Angie: *Okay, and he's not the most animated guy. When he listens, he's a good listener?*

Imants: *Mmm hmm.*

Angie: *He tries to be very present as a listener.*

Imants: *Yes, very present.*

Angie: *Is he a defenseman or are you?*

Imants: *He is a defenseman.*

Angie: *Is he a Boston fan?*

Imants: *Bruins?*

(This is a reference to the Boston Bruins National Hockey League team.)

Angie: *Yeah. He's showing me the Boston-Montreal*

Imants: *Series?*

Angie: *Yeah.*

Imants: *He went to school in Montreal.*

Angie: *Okay. I'm trying to figure out. I'm not sure if I was supposed to say Boston or Montreal.*

Imants: *Oh yeah, he probably would have been a big Habs fan.*

(The Montreal Canadiens National Hockey team is often referred to as the 'Habitants' or 'Habs'.)

Angie: *Okay, 'cause he's talking about the last series that was just up and commenting on the outcome.*

(That would be the Stanley Cup playoff series between the Montreal Canadiens and the Boston Bruins. So, it looks as though 'Canadian' could have been a reference to the hockey team and not the beer.)

Angie: *So he's commenting on that. And, is there, okay, I've never been to Montreal. Is there a 'Lafal' or 'Laval' or a place that sounds like that? Is Lafal or Laval like a city, or a place or a neighborhood?*

Imants: *I think it's a place, yeah. I think it's a university called Laval, but there's also a district.*

Angie: *Okay, yeah, because this feels like a neighbourhood, or an area, or a city suburb. You know what I mean? And there's like brown stones in this area, almost like New York to me. He's comparing, well I have no frame of reference, so I'm getting the upper west side of New York. So brownstones, and corner cafes, and that kind of neighborhoody kind of feel.*

Imants: *He was a neighborhoody kind of guy.*

Angie: *Is he really like nonchalantly, attitude, sarcasm a little bit?*

Imants: *Yeah.*

Angie: *In a really kinda, I don't know, he's really kind of: 'It's all bliss and harps', he says. I don't even know what to think of this guy he's so funny. He's like: 'There's little pink bunnies, you know; there's unicorns, Ferris wheels, and merry-go-rounds'. He's really a piece of work. Oh brother.*

Imants: *That would be him all right.*

Angie: *But what a shitty answer and what a good one. He's basically like 'I'm not gonna tell ya!'*

Imants: *It's all bliss and harps. Well, you know. Try telling someone who's alive what it's like.*

Angie: *Yeah. But I think the pink bunnies is funny. 'You can pretty much do what you want, and I certainly feel better', he says. He plays* *defense behind you, you're a winger?*

Imants: *Yeah.*

Angie: *Does he play defense behind you on the same side?*

Imants: *Usually. They'd switch whichever side they'd end up on.*

Angie: *Because I feel like there's trouble replacing him.*

Imants: *Yeah, he's good.*

Angie: *Well, that's what he's commenting on.*

Imants: *Yeah, right.*

Angie: *Trouble replacing him.*

This last interchange is a joke. It is true that my friend played hockey very well. But we play for fun so it really does not matter how well anyone plays. Yet he is joking here about how he cannot be replaced.

Angie seemed to have captured the spirit of my friend pretty well. But her reference to 'Laval' seemed off the mark. I knew that my friend had not gone to Laval University and I really doubted that he would have lived in the Montreal suburb of Laval. So what was the Laval reference? When I checked with his girlfriend, it turned out that my friend had lived on the corner of Laval Street and the Carre Saint-Louis in Montreal. That neighborhood, from what I could learn about it, matched the description given by Angie.

So what are we to make of this? Angie produced lots of information that was correct. And she produced specific information that turned out to be correct, but that neither of us knew at the time. I have since then heard her get various convoluted foreign-language names right. So we know that she produces lots of correct information and I have seen students get quite rattled when she is 'hot'. In formal research conducted at the University of Arizona at Tucson, my colleague Gary Schwartz found that over 80 per cent of the information produced by mediums during one part of his study was correct.[42] So we know that some mediums can provide correct information.

The problem is that that does not mean that the information is coming from dead people. Or other entities or beings such as angels, fairies, extra-terrestrials, demons, or whatever. Mediums could just be good at extracting that information from wherever it resides. Or, as Angie has said to my students, 'Some days I just think I'm a really good guesser'. However, the sense that a medium often has is that of interacting with a deceased person. I gave a bit of the details of the interchange between Angie and I in order to convey to the reader the impression that Angie appeared to be interacting with someone as though she were conversing with him. But such impressions of interaction do not necessarily mean that a medium is interacting with some independently objective being. That could just be the way in which our brains structure the information that we are remote viewing.

I tell dead people to get lost

Let me switch from a third-person to first-person perspective. I, too, have had some experiences in which I have had impressions of dead people. And not just dead people. I will use terms such as 'phantoms' and 'astral entities' to refer to whatever it is that is 'out there' that we cannot ordinarily perceive with our physical senses. I do not know whether there is anything 'out there' but then I am less and less convinced that there is anything 'in here' either, so let us just go along for a moment with the notion that there could be 'entities out there.' It seems appropriate to share three examples at this point, all of which help to dispel some of the glamor associated with mediumship.

Many years ago I led a weekly meditation group. We were a small but committed group, at least for a while, until we broke up. At one point, one of the members of my group told me that she would be dropping out. From what I recall, that had something to do with her 'spirit guides'. As she started talking about her spirit guides, I had the impression in my imagination of a hole in a floor around which a group of beings was hovering telling her various things. I described what these beings looked like. 'Those are my spirit guides', she told me. To which I replied something along the lines of the following. 'Well, they're not very smart. They don't actually know anything. They're just astral goons taking advantage of the fact that you can see them. I strongly recommend that you tell them to get lost'.

What I have found is that 'spiritual' people often feel that whatever phantoms show up in their lives must be 'spirit guides' and then dutifully devote themselves to doing whatever such phantoms seem to be telling them to do. I interpret the hole in the floor in my image as a representation of whatever it was in this woman's psyche that allowed for whatever source of information was out there to influence her if, indeed, there were any external influences in this case. The 'astral goons' appeared to be self-determining beings of some sort, possibly discarnate people wandering around bothering whichever living person would let them do so.

I am not sure what to make of this or whether my interpretation was correct. I managed to describe the 'spirit guides' to the satisfaction of the participant in my meditation group although, from what I recall, my characterization of these beings was quite vague. Note that my impression that these 'spirit guides' were not particularly well-behaved is consistent with the first example in which Angela's mother is told to 'get the Ouija board, or else'. I think that one lesson that we can take away from these sorts of apparent interactions is not to be intimidated by whatever it is that appears to be intruding in our lives, but to retain a sense of integrity.

The second example, which had a similar hole-in-the-floor theme, occurred a number of years later. A woman showed up in my office. She was with a spiritualist group whose members encouraged the deceased to communicate through them. Indeed, she said that she had started channeling deceased spirits by sitting at her typewriter at night and typing automatically. What was coming through was a book about what was wrong with the world and how it could be fixed. She told me that her guides had told her that I would find a publisher for the book. She looked tired and out of sorts. She was a single mother who worked during the day and stayed up late at night to write this channeled book. It was taking its toll on her.

When I looked at the manuscript of her book, it consisted of the message that the world was in a terrible state. But there was a solution, and the solution, which apparently we could not figure out for ourselves but needed discarnate spirits to tell us, was love. Furthermore, this woman who was doing the channeling was the embodiment of love and was going to bring these changes to the world. And all of that, without much more detail than that, was repeated over and over again using really bad grammar.

I told the woman that this was junk that no one would publish and that she should stop the automatic typing. When I got home that night I sat down to meditate and had the impression of a hole in the floor with beings who were pushing their agenda on this woman. My impression was that the automatic writing was not just the product of her own psyche, but that this woman had succeeded in opening herself to astral goons who were busy manipulating her. In my mind, I told them in no uncertain terms to get lost. The woman came back several months later to thank me. She had stopped the automatic writing, gone back to college, and looked healthy.

A third example comes from a dinner at a restaurant I had with some friends recently. We had been in a grocery store prior to going to the restaurant and one of my companions said that a discarnate person had attached to her between the cash register and the doors at the grocery store. I asked if it had been male and she said yes. I asked if he had a beard and she said yes. I asked if the beard was white and my companion said that she did not have a visual impression but that she could just feel the energy. I said that he had slurred speech. My impression was that he was a bit tipsy. My companion said that dead men attach themselves to her seeking her mothering energy. I said that that was what was happening in this case. And as we confirmed it, I 'saw' that the guy could see that I could 'see' him and he left. I said 'He just left'. I asked my companion if it felt as though he were gone and she said 'yes'.

Now, I do not know what happened in this case. I did not have a strong visual impression, more just a 'knowing'. This would be the difference between *clairvoyance* and *clairsentience.* Whereas the former is extrasensory visual perception, the latter is directly knowing something that we would not ordinarily know. I did, however, have impressions of the person's visual appearance that were partially 'clairvoyant'. Was this just *folie à deux*? Or is this the form that perception takes when we are looking at dead people? At any rate, I think that what this example reveals is the possibility that we could be interacting with dead people on an ongoing basis as we go about living our lives.

Past-life regression

Thus far in this chapter, I have been focusing on ostensible communication with the dead. But here is the thing. If we live forever after we die, then it would be temporally symmetrical if we also lived forever before we were born. That just makes sense. And I have had students who have told me not only that they are convinced that they have lived previously, but who claim that they can remember what happened in previous lifetimes. Of course they could be mistaken. And we could be mistaken to think that this is the only time we have been alive. Has there been any scientific research to settle this matter? The answer is, yes.

Ian Stevenson, a professor of psychiatry at the University of Virginia, spent many years in Asia examining cases of children who appeared to recall previous lives. My favorite are the cases in which children's birth marks or birth defects correspond to the manner of death of the previous personality that the child appears to have been. For instance, a boy who identified with a miller who had died after being struck on the head with a flour shovel had been born with a depressed area of the skull.[43] There are also North American cases that have been studied by others.[44] Stevenson did not think that there were any persuasive adult cases. I thought that there were a handful of those,[6] but that they were, overall, not as compelling as the child cases.

Adult cases are usually obtained by 'regressing' someone to a 'past life' using a guided imagery technique that could be labeled as 'hypnosis' by the person doing the leading. The technique that I have used is a guided imagery technique that is called the *christos technique*.[45-47] I have not done this sort of thing frequently, but when I have, the person I have regressed has always produced spontaneous past-life imagery. I myself have been regressed probably about ten times. I have found incarnations as a Viking, a Templar, a musician, a sailor, a World War II fighter pilot, a yeti. Yes. I used to be an abominable snowman who lived in a cave near a village. Sometimes I feel as though I have not changed much. I actually came across the World War II fighter pilot in my ordinary waking state by searching backwards in time for the origins of something that was going on with me. I no longer recall what that was, just the termination in the life of a fighter pilot.

But the more of these 'incarnations' I found, the less confident I became that any of them were actual lives that had been lived. I came to think of these scenarios as projections of whatever was going on in my psyche at the time. In that sense, it could be helpful for a person to experience such past-life scenarios. Indeed, one of my thesis students looked at the potential benefits of inducing past-life imagery. She did not find any benefits compared to present-life or future-life imagery, but that could have been because there was nothing psychologically wrong with the participants in her study.[12] But for all that past-life regression could be therapeutic, it need not say anything about whether or not we have lived before. Even if we find that the information during a past-life regression turns out to be correct, this does not mean that the person living now in some sense *was* that previous personality. This is analogous to the problem with mediumship. The medium churns out correct information, but is it being produced by a cognizant being in some other realm of reality?

A careful examination of the evidence, particularly from children's cases, strongly suggests that in some sense we have lived previously and will likely live again in the future. But closer examination of these cases raises questions of the sense in which one's identity is retained across incarnations. In what sense is it *I* who reincarnates? What is left without the body?

Disidentification

One approach to the survival hypothesis is to seek to determine for oneself whether one's identity is tied to the body. This was the approach taken by Venkataraman later known as Ramana Maharshi. On August 29, 1896 he felt as though he were going to die, so he lay down on the ground and imagined himself to be a corpse. Upon doing so he realized that his personality, including his sense of 'I', was unaffected by the 'death' of the body. According to Venkataraman, this realization was not the result of a deliberate reasoning process.[48]

The night of March 2, 2008, after contemplating Ramana Maharshi's experience with imagining himself to be dead, I had the following dream:

> ***Dream:*** *I had set in motion a process of bringing myself to the brink of death in order to have some insights just before actually dying, with the idea of bringing those back to my ordinary state. This process had progressed for a while when I came out of my bedroom*

> *in the modest house in which I was living and felt woozy. I thought, ok, this is it. I couldn't stand, so I decided I would lie down so that I could see the setting sun coming in through the kitchen window. That seemed soothing to me. A couple of my friends knew that I was doing this experiment and I imagined them finding my dead body on the kitchen floor. I figured that that was as good a place for them to find it as any. As I lay down on the floor, the next thing I knew I actually woke up on the couch in my living room.*

It was 7:00 a.m., half an hour before the time for which my alarm was set. I thought I would try to reinsert myself into the dream, but as I did so, I found that I could not identify with my 'body' on the kitchen floor but remained a sort of ghost hovering over it. Then my mother came down and started puttering in the kitchen and I got up.

In this case, dying in the dream resulted in actually waking up. I found that to be an interesting experience. It does not provide evidence for life after death, but it does give an illustration of what that experience could be like.

So where does that leave us with regard to the survival hypothesis? First, I think that ITC research should be developed further to see what is possible. It would be interesting if communication with the deceased became as commonplace as communication with the living through the use of telephones, texting, and e-mail. At that point the fear of death would likely dissipate. Second, more attention could be paid to the investigation of good mediums in order to better document the fact that they get correct information. There is still considerable denial in the conventional scientific community that such anomalous information transfer can occur. The more difficult question is whether there are ways of establishing if the information that mediums receive is actually coming from dead people or whether it is coming from physically existent sources through anomalous means. Third, there also needs to be more research into past and future incarnations along with states of being that exist between such states. In congruence with what I said previously about the nature of time, such states could be present in a dynamic now so that actions taken in an occurrent lifetime could affect both past and future incarnations. Are we essentially non-physical beings adopting a variety of temporally displaced personas? Is the lifetime we are experiencing just one of a number of projections into a space-time stream? An obvious approach would be to examine the experiences of people who appear to be able to 'remember' such lifetimes. We could also see if we could access aspects of the psyche, perhaps through the use of guided imagery techniques, in order to draw forth such information, and then check it against available documents.

A discussion such as this raises the more general question of our identity. Who are we? Can we answer that question for ourselves in a manner such as that of Ramana Maharshi? Is death just an awakening into another reality? As I said at the outset, none of what I have said provides decisive evidence for life after death, but is certainly consistent with such a notion. My own ideas have shifted as a result of these experiences so that I now think that survival is likely.

Chapter 6

Epilogue

I think that I have 'kept talking' long enough for now. For those who are still listening, I hope that what I have said has provoked some creative thinking about the nature of reality. What I have said in this book is at the edge of my understanding and I have tried to just lay it out as plainly as I can. I have tried to show how reality is more interesting than we usually suppose that it is by describing some of the experiences that I have had that have led me to the more expanded view of reality that I now have.

I think that science needs to step into the gap between our conventional interpretations of reality and those discussed in this book to check out the things that I have said. It is not difficult to do the relevant research. We just need to devote the resources to get it done. Elsewhere I have examined the practice of science and the need to restore integrity to scientific exploration so that research can be carried out effectively in areas of investigation such as those that I have described in this book.[2,14,17]

What about some of the more outlandish contentions, such as Jeanette's conviction that I took on someone else's cancer? Irrespective of whether or not that occurred for me, is that sort of thing even possible? When I started asking around, I was told several stories in confidence in which someone's illness disappeared only to reappear in the person who had healed her or who had asked to have the disease instead of her. When I had my research assistant search the medical and anthropological literature for such cases, she found nothing. Why do such cases not appear in the scientific record, if only to offer conventional explanations for them? If those of us who are engaged in healing can 'track' the diseases of others by mimicking apparent symptoms in our bodies, could we not also go one step further and relocate the disease to our own bodies? If the constituent parts of our bodies were to be just morphic fields expressing information, then why not? This is not to say that I think that the transmigration of disease does occur. Even in light of my own experience, I do not know. This is where science could step in to transmute ignorance into knowledge.

I have also tried to indicate how we have psychological resources available to us that we can use for improving the quality of our lives. This is a practical matter and not just a topic for speculation. I have tried to show how I have made use of knowledge from my dreams in order to guide my actions so as to be more successful. Having said that, I should clarify that I am not living in some sort of utopian world. I would like that, but that is not my current reality. And while I might be able to jiggle other people's energy levels a bit and make them feel as though they were living in a utopian world for a few days, I cannot magically make all of their problems go away either. In other words, drawing on our potential resources to improve our lives does not mean that all of our problems disappear. That is not to say that such end-states of exceptional well-being could not be achieved. I just want to make it clear that there are intermediate stages of being between abject suffering and blissful liberation.

Awakening from the dream

I have talked about precognitive dreaming, remote healing, the ability of dreams to shed light on a health crisis, and the survival hypothesis. These sorts of phenomena seem impossible. Of course, what we think is possible and impossible hinges on what we think reality is like. The more I scrutinize reality, the more it seems to me that the waking state of consciousness is more similar than dissimilar to the dream state and that many of the same rules apply. So just as we can wake up from a dream to realize its true nature, in the same way perhaps we can wake up from the ordinary waking state. In the spirit of this book, let me close with a dream from the night of November 24, 2007 that speaks to that possibility:

> ***Dreams:*** *The mood was melancholy. Various things were happening, but then I walked down on a university campus. There was a large cafeteria or meeting place on the ground floor, but I didn't go in there. I kept going to the left of it and ended up by a lake. It was cold out, but I was wearing shorts and bare feet. I noticed someone else who was dressed for warm rather than cold weather, but I was proud that I was able to walk around with shorts and bare feet in the cold.*
>
> **Waking Dream:** *The dream continued, and I had 'woken up' in my dream. I had thought that there were circulation problems in my body, but then I looked at the fingers on my right hand, and the middle fingers were black. Some of the other fingers were also partially blackened. I had taped pieces of cloth-like bandages around them, so that there were about 6 or 7 of these covering each of the middle fingers. I realized that my circulation problem was worse than I had thought. I freaked out and realized that I would have to go to see my physician, even though I didn't think that she could do much to help.*
>
> **Waking dream (cont.):** *But then I felt that I had 'woken up' and I looked at my right hand and saw that my fingers were a nice fleshy color and ok. It was only later that it occurred to me that this, too, was actually part of my dream, even though I'm not entirely sure of that. I really did wake up soon after that, though.*

There are at least three levels of waking up from dreams and each time I 'awaken' the nature of my reality has shifted. I 'wake up' from a dream to realize that I have circulation problems. Then I 'wake up' to find the circulation problems gone. Then I seem to wake up again to realize that the prior awakenings were telling me something about waking up. And eventually I really wake up to my ordinary waking state. So now, my question is, is there another level of awaking? Can we wake up from the ordinary waking state?

There is certainly one sense in which that can occur. It occurred for me when I was a teenager. I was the epitome of the 'good boy' trying hard to do everything that I was supposed to do to conform to society's expectations for me. But rather than feeling as though I were a part of the social fabric, I found myself isolated from it. I had kept my part of the bargain by doing everything I was supposed to do. I mentioned previously the complete trust in allopathic physicians, for example. But society had not kept its part of the bargain. Why was I unnecessarily poisoned with antibiotics to the point where I got seriously ill? And I had no idea what the point of life was. Why was there no practical assistance for resolving my existential angst? I started to resent cultural mores whose purpose seemed to me to be to stifle genuine inquiry into human nature rather than to encourage it.

To meet the requirements of my program in my second year of Engineering Science I had to choose a course from the Faculty of Arts and Science. I was looking through the university calendar when I came across a course description with the salacious question: 'What is the reason for our existence?' It was a philosophy course about Martin Heidegger's *Being and Time*.[49] Finally, something that fit the bill. I signed up for the course.

I was two months into the school year when my life fell apart. I remember that it was sunny out. And very warm. A leaf or two still clung to a tree here and there, splendidly gold and red. It was the kind of perfect day that only a day in Indian summer can be like. I was supposed to choose my metallurgy project that day. I think I could have enjoyed a metallurgy project had it not been for my preoccupation with the meaning of life. I remember sitting on a desk in a classroom with *Being and Time* open on my lap. We thingify everything, Heidegger was saying. We see ourselves as objects. But we are not things. We are *Dasein*, projecting into the future. Being in the world, we understand it the way *das Man*, 'they', tells us to. We interpret the world the way we are supposed to interpret it. And in so doing we fail to be authentic. We fail to grasp our own-most possibility, and thereby fail to experience love and joy.

I looked up. I was hot. We were in a basement room with a large window facing south and the sun was strong. A streetcar rattled by on College Street. The metallurgy project, we were told, must be approved. I closed my Heidegger, got up, and left the room. I walked to the administration office for my program and asked to transfer to the Faculty of Arts and Science. I felt like a naughty child who had done something wrong.

First, the administrators pulled out my file and looked at my academic record. I had received a substantial admissions scholarship to their program and had made honors standing in my first year in spite of the ten days I had spent in the hospital with my mysterious throat infection. Second, they gave me a pep talk. Just hang in there, they said. But I had made up my mind and was not about to be swayed. Third, the administrators got angry at me. We just don't want to see you end up sitting in a garbage can by the time you get out of school, they said. I could not believe that they would say that. Does everyone in a university program other than engineering end up sitting in a garbage can? At any rate, I realized that I would rather sit in a garbage can and know the meaning of life than send rockets into space or design computers and not know the meaning of life. It was only years later that I realized that engineering could have been as rich a context as any for resolving existential issues.

This was the breaking point. I became a full-time seeker. And Heidegger was the catalyst. He provided me with an understanding of the manner in which we allow ourselves to be bound to the expectations of 'they' that ensnare us in inauthenticity. In following my own convictions I had woken up and become authentic. And having awakened in this sense, it was impossible to fall asleep again. Nor would I ever wish to go back into the muddled stupor in which I had found myself prior to awakening to authenticity. It was also difficult to communicate with anyone who did not understand the distinction between inauthenticity and authenticity. Years later, I defined authenticity as acting on the basis of one's own understanding. I wrote about the nature of authenticity, analyzed the authentic and inauthentic modes of science, and showed how authentic scientific exploration was compatible with spiritual aspiration.[17]

But now the question arises of whether another awakening is possible. Are disease, old age, and death, along with a medical industry that seeks to ameliorate the suffering associated with those aspects of our lives, simply inauthentic modes of health? Is it possible to awaken to a life that is free of those afflictions? Do I need to rebel again against 'engineering' as I had done 40 years previously? Is that what the rant against the tattooed physician in my dreams was all about? The refusal to be part of the morphic field of human misery? Is this the second awakening in the dream of November 24, 2007 in which I see that there is no illness and that everything is actually ok? In other words, is an authentic state of uninterrupted health and equanimity possible?

Perhaps. It seems impossible to me that I could live a life of freedom from disease, old age, and death, and at least some dependence on conventional medicine. But then, at the start of my process, as a true believer of materialism, I also thought that precognitive dreams and remote influencing were impossible. Yet, over the years, I have slowly pushed back the boundaries of what is impossible until it has happened. So a little humility is appropriate. We become so convinced that we know what is possible and what is not that we fail to act even though it could be to our benefit to do so. At this point, the only thing of which I am convinced is my own ignorance. I do not know what is impossible anymore. But that opens the door to the anticipation that I could be surprised again by what nature could reveal. And then, perhaps you, the reader, and I can relish some version of reality as yet unimagined by us.

Notes

1. Wachowski A, Wachowski L. The Matrix [DVD]. USA: Warner Bros.; 1999.
2. Barušs I. The personal nature of notions of consciousness: a theoretical and empirical examination of the role of the personal in the understanding of consciousness. Lanham, Maryland: University Press of America; 1990.
3. Laplace PS. A philosophical essay on probabilities. New York: Dover; 1951, c 1814.
4. Barušs I. Characteristics of consciousness in collapse-type quantum mind theories. J Mind Behav. 2008;29(3):255-265.
5. Fuhrman J. Eat to live. Rev ed. New York: Little, Brown and Company; 2011.
6. Barušs I. Alterations of consciousness: an empirical analysis for social scientists. Washington, DC: American Psychological Association; 2003.
7. Smith DW, McIntyre R. Husserl and intentionality: a study of mind, meaning, and language. Dordrecht, Netherlands: D. Reidel; 1982.
8. Barušs I. Categorical modelling of Husserl's intentionality. Husserl Studies. 1989 Jan;6:25-41.
9. Barušs I. Beliefs about consciousness and reality: clarification of the confusion concerning consciousness. J. Conscious. Stud. 2008;15(10-11):277-292.
10. Barušs I, Moore RJ. Beliefs about consciousness and reality of participants at 'Tucson II'. J Conscious Stud. 1998;5(4):483-496.
11. Heintz L, Barušs I. Spirituality in late adulthood. Psychol Rep. 2001 Jun;88(3):651-654.
12. Woods K, Barušs I. Experimental test of possible psychological benefits of past-life regression. J Sci Explor. 2004;18(4):597-608.

13. Lukey N, Barušs I. Intelligence correlates of transcendent beliefs: a preliminary study. Imagin Cogn Pers. 2005;24(3):259-270.
14. Barušs I. Science as a spiritual practice. Exeter, UK: Imprint Academic; 2007.
15. Bartlett R. Matrix energetics: the science and art of transformation. New York: Atria; 2007. Quotations from pages 51 and 79.
16. Baker D. Esoteric healing. Essendon, Hertfordshire, UK: Douglas Baker; 1975-1978. 3 vol.
17. Barušs I. Authentic knowing: the convergence of science and spiritual aspiration. West Lafayette, Indiana: Purdue University Press; 1996.
18. Byrd RC. Positive therapeutic effects of intercessory prayer in a coronary care unit population. South Med J. 1988 Jul;81(7):826-829. Quotation from page 826.
19. Leibovici L. Effects of remote, retroactive intercessory prayer on outcomes in patients with bloodstream infection: randomised controlled trial. Brit Med J. 2001 Dec;323(7327):1450-1451. Quotation from page 1450.
20. Sheldrake R. The presence of the past: morphic resonance and the habits of nature. New York: Times Books; 1988.
21. Bartlett R. The physics of miracles: tapping into the field of consciousness potential. New York: Atria; 2009. Quotations from pages 80 and 48.
22. Merrell-Wolff F. Franklin Merrell-Wolff's experience and philosophy: a personal record of transformation and a discussion of transcendental consciousness. Albany, NY: State University of New York Press; 1994.
23. Merrell-Wolff F. Transformations in consciousness: the metaphysics and epistemology. Albany, NY: State University of New York Press; 1995.
24. Merrell-Wolff F. Mathematics, philosophy & yoga: a lecture series presented at the Los Olivos Conference Room in

Phoenix, Arizona, in 1966. Phoenix, AZ: Phoenix Philosophical Press; 1995.

25. Assagioli R. Psychosynthesis: a manual of principles and techniques. New York: Penguin; 1965.
26. Broughton RS. Parapsychology: the controversial science. New York: Ballantine; 1991.
27. Bengston WF, Krinsley D. The effect of the "laying on of hands" on transplanted breast cancer in mice. J Sci Explor. 2000;14(3):353-364.
28. Bengston W. The energy cure: unraveling the mystery of hands-on healing. Boulder, Colorado: Sounds True; 2010.
29. Schwartz SA, Dossey L. Nonlocality, intention, and observer effects in healing studies: Laying a foundation for the future. Explor. 2010 Sept;6(5):295-307.
30. Swanson C. Life force: the scientific basis: breakthrough physics of energy medicine, healing, chi and quantum consciousness. Tucson, Arizona: Poseidia Press; 2010.
31. Barušs I. Speculations about the direct effects of intention on physical manifestation. J Cosmol. 2009 Dec; 3: 590-599.
32. Office of Dietary Supplements (US). Dietary supplement fact sheet: carnitine [Internet]. Bethesda, MD: National Institutes of Health (US); 2006 June [updated 2006 June 15; cited 2012 Aug 26]. Available from: http://ods.od.nih.gov//factsheets/Carnitine-HealthProfessional/
33. Paulson D, Bowen ME, Lichtenberg PA. Successful aging and longevity in older old women: the role of depression and cognition. J Aging Res. 2011 May;2011:1-7.
34. Langer EJ. Counter clockwise: mindful health and the power of possibility. New York: Ballantine; 2009.
35. Alesci S, Manoli I, Costello R, Coates P, Gold PW, Chrousos GP, et al, editors. Carnitine: the science behind a conditionally essential nutrient. Ann N Y Acad Sci. 2004 Nov;1033. Quotation from page x.
36. Certo M, Franca M, Gomes M, Machado R. Liver teratoma.

Acta Gastroenterol Belg. 2008 Apr/Jun;71(2):275-279.
37. Baker D. The spiritual diary. Potters Bar, UK: College of Spiritual Enlightenment and Esoteric Knowledge; 1977.
38. Irwin HJ. An introduction to parapsychology. 2nd ed. Jefferson, NC: McFarland & Company; 1994.
39. Braude SE. Immortal remains: the evidence for life after death. Lanham, Maryland: Rowman & Littlefield; 2003.
40. Barušs I. Failure to replicate electronic voice phenomenon. J Sci Explor. 2001;15(3):355-367.
41. Barušs I. An experimental test of instrumental transcommunication. J Sci Explor. 2007;21(1):89-98.
42. Schwartz GE. The afterlife experiments: breakthrough scientific evidence of life after death. New York: Pocket; 2002.
43. Stevenson I. Reincarnation and biology: a contribution to the etiology of birthmarks and birth defects. Westport, CT: Praeger; 1997.
44. Bowman C. Return from heaven: beloved relatives reincarnated within your family. New York: HarperCollins; 2001.
45. Glaskin GM. Windows of the mind: discovering your past and future lives through massage and mental exercise. New York: Delacorte; 1974.
46. Glaskin GM. Worlds within: probing the Christos experience. London, England: Wildwood House; 1976.
47. Glaskin GM. A door to eternity: proving the Christos experience. London, England: Wildwood House; 1979.
48. Mahadevan TMP. Ramana Maharshi: the sage of Arunacala. London: George Allen & Unwin; 1977.
49. Heidegger M. Being and time. Macquarrie J, Robinson E, translators. New York: Harper & Row; 1962, c 1927.

Iff Books is interested in ideas and reasoning. It publishes material on science, philosophy and law. Iff Books aims to work with authors and titles that augment our understanding of the human condition, society and civilisation, and the world or universe in which we live.

9781780995458